中国政府无偿援助蒙古国文化遗产项目
Gratis assistance project over Mongolia's cultural heritage from Chinese government

博格达汗宫博物馆维修工程

The Maintenance Project of Bogd Khaan Palace Museum

БОГД ХААНЫ ОРДОН МУЗЕЙН СЭРГЭЭН ЗАСВАРЛАЛТЫН АЖИЛ

陕西省文物保护研究院 著
Shaanxi Institute for the Preservation of Cultural Heritage

博格达汗宫博物馆维修工程

The Maintenance Project of Bogd Khaan Palace Museum

编辑委员会 Editorial Board

目　录　　TABLE OF CONTENTS

序

2005年，中国政府为了进一步推动中蒙的文化交流，国家文化部、国家文物局与蒙古国教育文化科学部签署协议，确定无偿援助蒙古国文化遗产保护项目——博格达汗宫博物馆维修工程。这是中蒙两国在文化遗产保护领域进行的首次合作。国家文物局考虑后将此任务交由陕西省文物保护研究院（原西安文物保护修复中心）承担实施，这是国家文物局首次将援外文物保护项目交由省级文物保护机构承担。

陕西省作为具有丰富历史文化资源的文物大省，文物工作得到历届陕西省委、省政府高度重视与倾力支持，文物保护事业不仅取得了丰硕成果，而且文物机构建设、专业人才队伍培养、创新保护模式得到令人瞩目的发展。尤其在中外文物保护的合作与交流方面，陕西起步早，成效显著。合作的方式也由初期单一引进发达国家的先进技术和仪器设备等，发展到方法理念的学术交流与项目合作，逐渐步入了承担涉外考古、援外工程和技术输出的新阶段。

陕西省文物保护研究院是陕西省政府批准成立的陕西省第一个文物保护专业机构，二十多年来，先后承担了国家科技部、国家文物局和陕西省文物局十多项重大文物保护科研课题，获国家和省级科研奖项七项，申报技术专利五项。被国家文物局评为“汶川地震灾后文物抢救保护工作先进集体”。目前拥有“砖石质文物保护国家文物局重点科研基地”、科技部在全国唯一设立的一处“文物保护国际科技合作基地”，陕西省委首批 “三秦学者”设岗单位。并在与德国、日本、意大利、美国等开展的文物保护科技的中外合作方面卓有成效，有力地推动了陕西省文保科技的发展，是国内外具较高知名度和影响力的综合文物保护机构。

援助项目博格达汗宫博物馆位于蒙古国首都乌兰巴托市南郊图拉河畔，始建于1893年，曾是蒙古国最高宗教领袖八世博格达・哲布尊丹巴居住和进行政教活动的场所。建筑借鉴了古代汉式传统工程做法，具有蒙藏文化风格，是蒙古国最重要的历史文化遗产之一。鉴于对外援助项目的重要性和特殊性，陕西省文物局高度重视，局领导和相关处室协调成立专门项目组，抽调精明强干的专业技术队伍，短时间内高效完成了保护设计方案。经中国国家文物局专家评审，提交蒙古国审核通过。

2006年5月27日，博格达汗宫博物馆维修工程正式开工，总投资600万元人民币。蒙古国总理恩赫包勒德为工程开工剪彩，国家文物局、中国驻蒙大使馆、陕西省文物局等相关领导出席了开工典礼。工程实施历时十七个月，圆满完成了大门、牌楼、宫墙、照壁、栅栏等10个单体的古建筑保护维修。2007年10月工程竣工，由中蒙文物专家进行了联合验收，认为依据充分，技术合理，质量优良，保存了文物的真实性和完整性，符合国际文物保护修复原则，同时对中国文物工作者科学严谨的敬业精神和高水平的技术能力给予高度赞誉。

2010年12月，陕西文物保护研究院派出工作组赴蒙对援助项目进行了质量回访。经现场核查，历经蒙古国三年多极端气候的考验工程质量保持良好。博格达汗宫博物馆维修工程能得到中外各方的充分肯定，首先是基于陕西文物工作者的科学态度。项目实施中陕西文物保护研究院利用科技人才优势，将科研与项目实施相结合，在文物环境、材料、工艺、病害全面调查基础上，对彩绘颜料、建筑材料、工艺配比、土壤成分等6大类67种样进行科学检测分析，完成多项关键技术研究，形成了中英文《蒙古国博格达汗宫建筑彩绘保护修复研究报告》，提出了科学合理的保护技术措施，为工程项目实施提供了可靠依据。其次是陕西文物工作者精湛的修复水平。工程中我们将中国文物保护理念与国际古迹保护准则相结合，既严格遵循最小干预的国际原则，保持蒙藏古建筑的原真性，同时鉴于博格达汗宫与我国山西古建筑工艺的同源性，在修复方法上运用中国文物保护理念，又完整地保留了蒙古传统古建筑特色。既达到了保护文化遗产的目的，也符合蒙古民众对维修后的传统审美要求，良好地展现了历史文化遗产的美好形象。三是陕西文物工作者对文化遗产高度负责的态度和使命感。在项目实施期间，陕西省文物保护研究院着眼于文化遗产长期的保护与管理，结合现场实践，义务培训了近10位蒙古国文物保护工作人员，创造条件安排4人次赴中国陕西进行为期三个月的系统学习，基本掌握了文化遗产保护技术、古建筑保护管理知识，为博格达汗宫门前区保护项目完成后的日常管护提供了专业人员保障。博格达汗宫博物馆维修工程对陕西省文物工作者来讲，是一次走出国门磨练队伍的良机，同时是充分展示陕西省文物工作者敬业精神、专业风采的平台。陕西文物保护研究院克服援外文物保护工程的诸多困难，如当地建筑材料缺乏、跨国运输周期长、入关程序繁复、施工气候条件差等问题。以科学严谨、精明强干、专业管理、优质高效赢得了受援国的高度赞誉。整个施工期间蒙古国各界人士非常关注，各类媒体全方位大量持续的报道，树立了中国文物工作者的良好形象。同期在蒙的日本、德国、瑞士、美国等多位国际古迹保护机构专家学者专赴工地考察，同我方进行了技术交流，表达了希望与我方合作的意愿。

蒙古国文物保护援助项目的圆满完成具有标志性意义，其意义已超出了工程本身内涵，这是陕西省文物保护专业技术队伍走出国门的成功国际实践，更是陕西省文物保护综合实力在国际文化遗产保护领域的一次集中展示。体现了陕西从文物资源大省迈向文物保护强省的转变；进一步证明了陕西文物工作是提高陕西在国际影响力的积极力量，更是文物工作服务大局，提升中国文化软实力和展示国家形象的有效手段。

本书翔实记录了博格达汗宫博物馆维修工程的修缮过程，既是对过去工作的一个总结，也是今后工作的一个新起点。希望陕西文物工作者再接再厉，以新的成果回报社会。

感谢所有关心支持参与蒙古国博格达汗宫门前区保护项目的领导、专家学者和工作人员。

陕西省文物局局长：赵　荣

2013年6月2日

Foreword

For further promoting the cultural exchange between People's Republic of China and Republic Mongolia, Ministry of Culture of the P.R.C. along with State Administration of Cultural heritage signed an agreement with Mongolian Ministry of Education and Culture in 2005, in which Chinese Government Authorities agreed to gratis assist the maintenance project on Mongolian Bogd Khaan Palace Museum. This is the first cooperation between China and Mongolia in the field of cultural heritage protection and conservation. State Administration of Cultural heritage authorized Shaanxi Institute for the Preservation of Cultural Heritage (originally called Xi' an Center of Conservation and Restoration of Cultural Heritage) to carry out this project, this is also the first time that State Administration of Cultural Heritage hand over an international-level project on cultural heritage protection to a provincial institution.

As a Province who possesses abundant historical and cultural resources, works concerned about historical heritage in Shaanxi Province have received highly attentions and supports from Shaanxi Provincial Party Committee and Government, which yielded fruitful results in institution construction, personnel training as well as innovative mode of conservation and restoration. Especially in the field of transnational cooperation in historical heritage protection, Shaanxi Province not only gains a early origin but also gets a remarkable achievements. As to the approaches of cooperation, it shows a changing process from simply importing advance technics and instruments from developed countries to mutual exchanging in theory and methodology, and gradually starts to undertake works overseas and exports new technology of our own.

Shaanxi Institute for the Preservation of Cultural Heritage is a first professional institution on historical heritage protection in Shaanxi Province approved by Shaanxi Provincial Government, and have undertaken tens of significant research projects from Ministry of Science and Technology of P.R.C., State Administration of Cultural Heritage and Shaanxi Provincial Bureau of Cultural Heritage within around twenty years, and has been awarded both state-level and provincial-level prizes for seven times, and also has declared five technical patents, and also has been honorarily named as "Advanced Group of Historical Heritage rescue and protection after Wenchuan Earthquake" by State Administration of Cultural Heritage. Shaanxi Institute for the Preservation of Cultural Heritage is a professional and comprehensive institute with relatively high fame and influence in both domestic and overseas, and so far, it possesses several titles as "Key Scientific Research Base on the Protection of Brick Masonry", "International Technology Cooperation Base on the Protection of Cultural Heritage" (exclusively launched by Ministry of Science and Technology of P.R.C.) as well as "San Qin Scholar" unit approved by Shaanxi Provincial Party Committee, and has also been crowned with success in cooperation with German, Japan, Italy, America, etc., with regards to historical heritage conservation, all of which contributed to the development of Shaanxi historical heritage conservation technology.

Bogd Khaan Palace Museum is located at the foot of Mountain Bogd Khaan, in the south outskirts of Ulaanbaatar, it was first built in 1893, the eighth Bogd Jivzundamba Hutukutu, or Bogd Khaan, the highest religious leader of Mongolia, once lived there and conducted religious activities. The architectures took the examples from the traditional practice in ancient Chinese style, and also bears a Mongolian and Tibetan style, it is one of the most important historical heritage in Mongolia. Since the overseas assistant project is very special and significant,

Shaanxi Provincial Bureau of Cultural Heritage paid highly attention to it and a special project team consisted of professional staffs was established by relevant authorities, and soon the project proposal effectively came out within a short period, and then experts from State Administration of Cultural Heritage reviewed this proposal and finally Republic Mongolia approved it.

On 27th May 2006, the maintenance project of Bogd Khaan Palace Museum officially commenced with the total investment fund of 6 million RMB. Mongolia Prime Minister Miegombyn Enkhbold attended the commencement ceremony and cut the ribbon, relevant leading cadres from State Administration of Cultural Heritage, Chinese Embassy in Mongolia as well as Shaanxi Provincial Bureau of Cultural Heritage attended this ceremony too. The maintenance project lasted for 17 months, during which the maintenance works on the front gate, decorated archway, palace wall, screen wall, stockade, etc. were smoothly completed. This project was successfully accomplished in October 2007, experts from both China and Mongolia co-examined it and expressed high praise on the works being done.

In December 2010, Shaanxi Institute for the Preservation of Cultural Heritage sent out its working team and revisited the project site for re-inspection, and through inspection, the quality of project is very satisfactory that the works being done are quite perfect after undergoing three years extreme weather test in Mongolia.

The reason why this maintenance project of Bogd Khaan Palace Museum received fully appreciate from parties of both China and Mongolia is that: Firstly, the scientific attitude of staffs and workers from Shaanxi Province. During this project, Shaanxi Institute for the Preservation of Cultural Heritage made good use of its professional advantages, on the basis of comprehensive investigation on environment, materials, technics and damages, they combined the academic research with practical situation, and conducted detection and analysis on 67 kinds of samples(broadly 6 categories) with respect to colored paintings, construction materials, technics as well as soil constituent, and also finished several researches on some key technologies and finally form a Research Report on maintenance of colored paintings of Bogd Khaan Palace Museum, which provided scientific technical measures and supports for implementing this project. Secondly, it is due to their professional skills on historical heritage protection that can receive much appreciate, they incorporated the Chinese idea on cultural heritage protection with international principles, not only had kept the original appearance of Mongolia-Tibetan style, but had also maintained the characteristics of Mongolia’s ancient architectures, which not only achieved the goal of protection but also fitted aesthetic requirement of Mongolian people. Thirdly, it must be their responsibility and sense of mission that contributed to the final success. During project implementing, and taking into account of the practical situation, Shaanxi Institute for the Preservation of Cultural Heritage provided training program for around 10 staffs from Mongolia, and created every possibilities to allow 4 people coming to Shaanxi for a 3-months systematic learning, after this period, those visiting people basically grasped the technology of protecting historical heritage which provided the guarantee for Bogd Khaan Palace Museum’s daily management.

The maintenance project of Bogd Khaan Palace Museum is a good chance for staffs who work on cultural heritage in Shaanxi Province to go abroad and gain experiences, and is also an opportunity from which they have showed their professional dedications. Shaanxi Institute for the Preservation of Cultural Heritage had conquered many difficulties during implementing this project, such as lack of constructing materials, long-term transnational transportation, complicated customs procedures as well as bad weather conditions, etc. and finally they received

highly praise from Mongolian authority. In the whole course of the project, people from all circles paid close attention and kinds of Media continuously reported this project. In the meantime, experts from Japan, Germany, Switzerland, America, etc. also came to the project spot and exchanged their thoughts with us expressing their wishes of cooperation.

The completion of this maintenance project is a landmark of transnational practice on cultural heritage protection, and is also a display of the comprehensive strength of Shaanxi Province in the field of international historical heritage conservation, which reflects the transformation of Shaanxi Province from a Province of enriched historical heritage resources to a Province who is capable of doing transnational historical heritage protection. Further more, it also improves the international influence of Shaanxi Province and reflects the good figure of our nation.

This book detailedly records the entire course of the maintenance project of Bogd Khaan Palace Museum, it is a summary of the past as well as a new start for the future. Hereby I sincerely hope that those who participated in this project could make persistent efforts and also make new contributions to the society hereafter.

Director of Shaanxi Provincial Bureau of Cultural Heritage *Rong ZHAO*

2 June 2013

博格达汉宫博物馆门前区
The Front Square of Bogd Khaan Palace Museum

重檐歇山式绿色铁皮瓦彩画大门
Green Iron Sheet Tile Colored Drawing Gate with Multipleeaves Xieshan style

歇山式绿色铁皮瓦彩画中牌楼
Green Iron Sheet Tile Colored Drawing Middle Archway with Xieshan style

歇山式绿色铁皮瓦彩画西牌楼
Green Iron Sheet Tile Colored Drawing West Archway with Xieshan style

歇山式绿色铁皮瓦彩画西牌楼侧面
Side Face of Green Iron Sheet Tile Colored Drawing West Archway with Xieshan style

旗杆
Flagpole

悬山式绿色铁皮瓦彩画东便门
East Side Door

悬山式绿色铁皮瓦彩画西便门
West Side Door

南宫墙
South Palace Wall

栅栏墙
Fence Wall

二龙戏珠砖雕照壁
Screen Wall with Tile Carving Patterns like Two dragons play with a pearl

彩画守护门神正面门扉
The Right Side Door Leaf with Colored Painting of Door-God

大门梁架局部
Beam Frame of the Front Gate

大门梁架局部
Beam Frame of the Front Gate

大门梁架局部
Beam Frame of the Front Gate

大门彩画局部
Colored Painting on the Front Gate

大门彩画局部
Colored Painting on West Subordinate Building of the Front Gate

大门彩画局部
Colored Painting on east side of the Front Gate

大门彩画局部
Colored Painting on the Front Gate

大门屋檐局部
Part of the Eave

大门侧面
Flank Side of the Front Door

中牌楼一角
A Corner of the Middle Archway

中牌楼彩画局部
Colored Painting on Middle Archway

西牌楼彩画局部
Colored Painting on West Archway

大门屋顶镀金脊饰件
Gold-plated Ridge Decoration

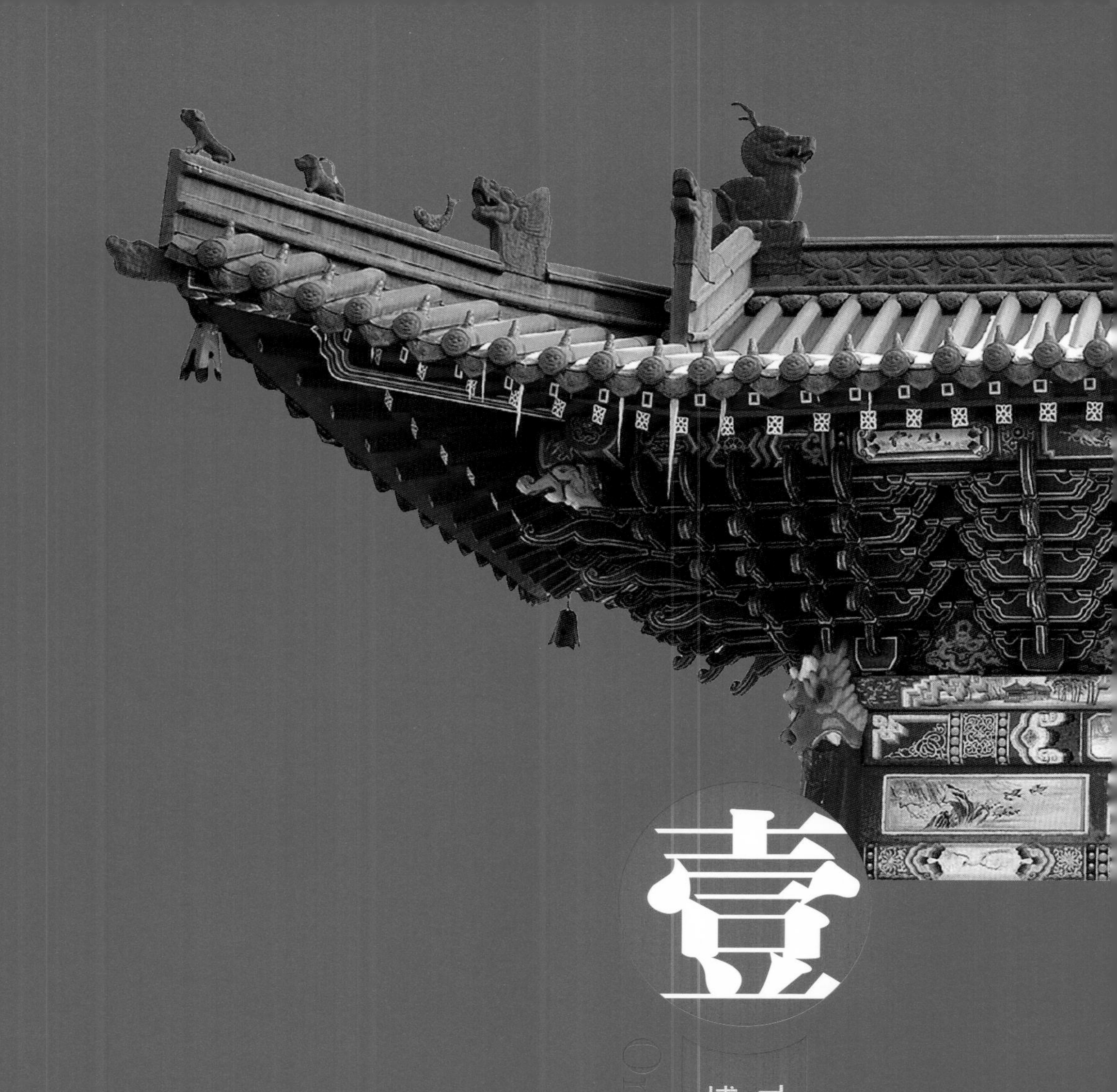

壹

One

前言

PREFACE

前言

PREFACE

The Maintenance Project of Bogd Khaan Palace Museum

博格达汗宫博物馆维修工程

1 前言

“博格达汗宫博物馆维修工程”是中国政府首次无偿援助蒙古国文化遗产保护项目，项目投资600万元人民币，2006年5月27日开工，2007年10月8日竣工。

博格达汗宫博物馆位于蒙古国首都乌兰巴托市南郊的博格达汗山下，始建于1893年，采用蒙藏文化风格与中国古代建筑相结合的方式建成，是蒙古国最高宗教领袖八世博格达·哲布尊丹巴居住和进行政教活动的场所，代表了当时建筑科学技术的水平，在蒙古地区的建筑史上占有重要地位并具有较高的历史研究价值，是蒙古国重要的历史古迹之一。1926年，博格达汗宫改建成国家博物馆。

2004年，中国文化部孙家正部长、国家文物局单霁翔局长访蒙，中蒙双方达成两国间文化交流协议，包括博格达汗宫博物馆门前区保护维修、合作考古和文物展览等三个项目。

2005年，经过中国文化部、国家文物局和蒙古国教育文化科技部的共同协商确定，中国政府无偿援助蒙古国文化遗产保护项目——“博格达汗宫博物馆维修工程”开始启动。中国国家文物局委托陕西省文物保护研究院负责该项目的具体实施。

在中国驻蒙大使馆、中国文化部、中国国家文物局、蒙古国教育文化科技部和陕西省文物局等各方的大力支持下，2006年5月至2007年10月，“博格达汗宫博物馆维修工程”取得圆满成功。经过中蒙文物专家组对维修工程进行的联合验收，认为该工程严格遵循国际公认的文物保护修复原则，所采取的技术措施合理，彩画保护修复依据可靠，较好地保存了文化遗产的真实性和完整性，工程质量和效果良好，一致通过验收。

作为中国与蒙古国两国间文化交流的重要组成部分，“博格达汗宫博物馆维修工程”是中国文化遗产保护对外援助的重点项目之一，也是陕西省文物系统首次走出国门承担的援助项目。“博格达汗宫博物馆门前区保护维修工程”的实施，不仅向蒙古国展示了中国文物保护工作者精湛的维修技术，更是中国文化软实力和综合国力不断提升的具体体现，是展示国家形象的有效手段。该援蒙工程的圆满完成，标志着陕西省文物保护研究院由文物保护技术国际受援单位，成长为代表中国文物保护水平走出国门的国际援助单位，赢得了国内外专家学者的赞誉，保护维修工程的质量和效果受到蒙古国政府、中国驻蒙大使馆、中国国家文物局等各方的高度赞誉，充分展示出中国文化遗产的科学保护理念和中国文物工作者的高超技术水平，同时也为中蒙两国文化遗产保护进一步交流合作打下了重要基础。

签字仪式
Signing Ceremony

1.Preface

“The maintenance project of Mongolia Bogd Khaan Palace Museum ” is a protection project over Mongolia cultural heritage gratis conducted for the first time by Chinese government with project funds of 6 millions RMB. The project successfully came into operation in May 27th 2006 and perfectly completed in Oct. 8th 2007.

Bogd Khaan Palace Museum is located at the foot of Mountain Bogd Khaan, in the south outskirts of Ulaanbaatar. It was first built in 1893 displaying a combination of Tibetan cultural style and Chinese ancient architectural pattern. The eighth Bogd Jivzundamba, the highest religious leader of Mongolia, once lived there and conducted religious activities. This palace, one of the most important historical sites in Mongolia, represents the architectural technology level of that time and plays a significant part in Mongolia' s architectural history with great research value. Bogd Khaan Palace was converted into a State Museum in April 1st 1926.

Sun Jiazheng, Chinese minister of the Ministry of Culture, and Shan Jixiang, director of State Administration of Cultural Heritage visited Mongolia in 2004. The two sides reached an agreement on Sino-Mongolia culture exchange which contained the maintenance work of the front square of Bogd Khaan Palace Museum, archaeological studies cooperation and cultural heritage exhibition.
By negotiation among Chinese Ministry of Culture, Chinese State Administration of Cultural Heritage and Mongolia Ministry of Education and Culture, " The front square maintenance project of Mongolia Bogd Khaan Palace Museum " , gratis assisted by Chinese government over Mongolia cultural heritage, initiated in 2005 . Xi'an Centre for the Conservation and Restoration of Cultural Heritage was specially commissioned by Chinese State Administration of Cultural Heritage for project implementation.

Under energetically support from Embassy of People' s Republic of China in Mongolia, Chinese Ministry of Culture, Chinese State Administration of Cultural Heritage, Mongolia Ministry of Education and Culture and Shaanxi Provincial Bureau of Cultural Heritage, etc., " The maintenance project of Mongolia Bogd Khaan Palace Museum" successfully came into operation in May 2006 and perfectly completed in Oct 2007. Chinese and Mongolian experts regarded this project as having strictly followed the international recognized principles of cultural heritage protection and maintenance after their acceptance check, the technology measures being taken were appropriate, the practice of colored paintings protection and maintenance was perfect which nicely preserved the authenticity and integrity of cultural heritages, the construction quality and effect were quite well which got unanimous approval.

" The maintenance project of Bogd Khaan Palace Museum" is one of Chinese most important projects on cultural heritages protection assistance to foreign country, which is also an assistance project which Shaanxi Provincial Bureau of Cultural Heritage undertaking overseas for the first time. The implementation of " The front square maintenance project of Bogd Khaan Palace Museum" is not only a manifestation of Chinese superb maintaining skill, but also a reflection of both Chinese booming cultural soft strength and comprehensive national strength, it is a effective way to show our national image. The successful completion of the project indicates the growth of Xi'an Centre for the Conservation and Restoration of Cultural Heritage from an international aid-receiving organization on cultural heritage protection to an international aid-supporting organization.Xi'an Centre for the Conservation and Restoration of Cultural Heritage, who gained recognitions and praises from experts and scholars domestically and overseas, showed Chinese capability on cultural heritage protection. The quality and effect of the maintenance project, which fully demonstrated Chinese scientific protection philosophy on cultural heritages and the superb technical skills from Chinese cultural heritage workers, had received highly praises from all parties including Mongolian government, embassy of People' s Republic of China in Mongolia and Chinese State Administration of Cultural Heritage. Meanwhile, a solid foundation had been established for further cooperation over Sino-Mongolia cultural heritage protection.

开工剪彩（左二蒙古国总理米耶贡布·恩赫包勒德、左三中国驻蒙大使高树茂）
Cutting Ribbon at Open Ceremony (The second from the left is Mongolian Prime Minister named Miet Kampot. Enkhbold, the third from the left is Chinese Ambassador in Mongolia named Gao Shumao)

中国国家文物局向蒙古国教育文化科技部赠送礼物
The Gift for Mongolia Ministry of Education and Culture from State Administration of Cultural Heritage

中国驻蒙大使余洪耀视察现场
Chinese Ambassador to Mongolia Yu Hongyao inspecting the Site

中国驻蒙大使馆文化参赞王大奇和一等秘书周晶视察现场
Cultural Counselor from Chinese Embassy in Mongolia Wang Daqi and First Secretary Zhou Jing inspecting the Site

竣工典礼
Completion Ceremony

维修纪念碑
Maintenance Project Monument

蒙古国向维修单位及个人颁发证书
Mongolia awarding Certificates to Maintaining Organization and Individual

ӨРГӨМЖЛӨЛ

БНХАУ-ЫН ШИ-АН ХОТЫН ТҮҮХ, СОЁЛЫН ДУРСГАЛТ ЗҮЙЛИЙГ СЭРГЭЭН ЗАСВАРЛАХ ТӨВИЙН ХАМТ ОЛОНД

Монгол улсын түүх, соёлын хосгүй үнэт дурсгал болох Богд хааны ордон музейн урд хэсэгт орших Энх-Амгалангийн хаалга, Асарт хаалга, Ямпай хаалга, урд талбайг мэргэжлийн өндөр түвшинд, технологийн дагуу сэргээн засварласан Бүгд Найрамдах Хятад Ард Улсын Ши-Ан хотын түүх, соёлын дурсгалт зүйлийг сэргээн засварлах төвийн хамт олонд гүнээ талархсанаа илэрхийлж энэхүү **Өргөмжлөл**-ийг өргөн барив.

МОНГОЛ УЛСЫН БОЛОВСРОЛ, СОЁЛ, ШИНЖЛЭХ УХААНЫ ЯАМ

УХА0148

2007. X. 08 Улаанбаатар хот

奖 状
西安文物保护修复中心全体工作人员：
对于贵单位参与修复的蒙古国一等文物博格达汗宫博物馆门前区（和平门，牌楼，照壁）所作出的杰出贡献，深表感谢，特颁发此奖状。
蒙古国文化科技教育部。
乌兰巴托市 2007年10月8日

Certificate of Honor
Staff at Xi' an Center for the Conservation and Restoration of Cultural Heritage:
We sincerely express our gratitude for your outstanding contribution over the front square maintenance project of Bogd Khaan Palace Museum (including Peace Gate, Decorated Archway and Screen wall), which is a first-class cultural heritage in Mongolia, and hereby deliver this Certificate.
Mongolia Ministry of Education and Culture
Ulaanbaatar Oct. 8th 2007

贰

Two

历史背景及建筑艺术

HISTORICAL BACKGROUND AND ARCHITECTURAL ART

历史背景及建筑艺术

HISTORICAL BACKGROUND AND ARCHITECTURAL ART

The Maintenance Project of Bogd Khaan Palace Museum

博格达汗宫博物馆维修工程

历史背景及建筑艺术

博格达汗宫博物馆修建于1893年，曾是蒙古最高宗教领袖八世博格达·哲布尊丹巴居住和进行政教活动的场所。1924年5月，第八世哲布尊丹巴圆寂。1926年4月1日建成博物馆，收藏了博格达汗使用过的生活物品、宗教法器、传统绘画和17至20世纪部分蒙古国的珍贵文物。

1.历史沿革

1869年，八世博格达在拉萨出生。1871年，他作为八世博格达·哲布尊丹巴的身份得到证实。1874年10月，博格达·哲布尊丹巴和他的父母兄妹一起离开西藏到达乌兰巴托，受到了当地贵族和崇拜者们隆重的欢迎，成为蒙古地区的宗教领袖。1911年清朝灭亡后，他作为博格达汗取得了政治大权，直至1924年去世。

博格达汗宫博物馆经历了晚清到民国及蒙古独立三个不同的政治变革过程，直到1924年第八世哲布尊丹巴圆寂。博格达汗宫博物馆目睹了漠北结束归附清王朝统治而独立建国前后的这段历史，作为这一历史的直接见证，具有补史、证史、正史的重要作用，具有重要的历史价值和研究价值。

2.地理位置

博格达汗宫博物馆位于蒙古国首都乌兰巴托市南郊博格达汗山脚下的图拉河畔之临山近水、地势平坦的川地上，有着优雅而风景如画的环境。

地理位置图
Geographic location

3.建筑布局

博格达汗宫博物馆现状平面，呈一“回”字形长方形的布局格局。横向分三路组合，中路为夏宫的主体，即原哲布尊丹巴呼图克图居住夏宫所在。左右两路在中路院落的东西两侧。宫门外为前庭广场及广场内设施——照壁、木牌楼等。广场有木栅栏设施。

中路院落建筑从前至后依次为：前庭广场、宫门、钟鼓亭、一进院、二进院、三进院。中路建筑中的二进院，是哲布尊丹巴呼图克图夏季所居宫殿的主体，俗称夏宫。

博格达汗宫博物馆左右两路院落，依外侧的木构围墙。木构围墙居中两侧各有通东西以外的便门一座。左侧靠南位置，有二层藏式建筑的冬宫一座，面西坐东朝向，此殿原为藏式二层构制，后经改建，门窗改为俄式风格，二层平顶也改为简易歇山式屋顶。

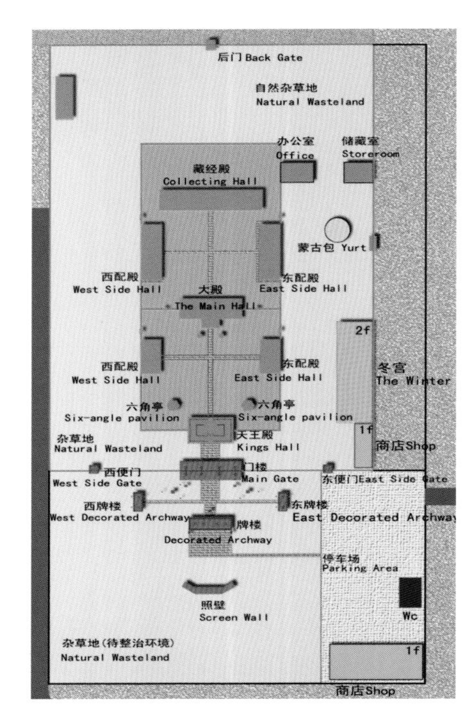

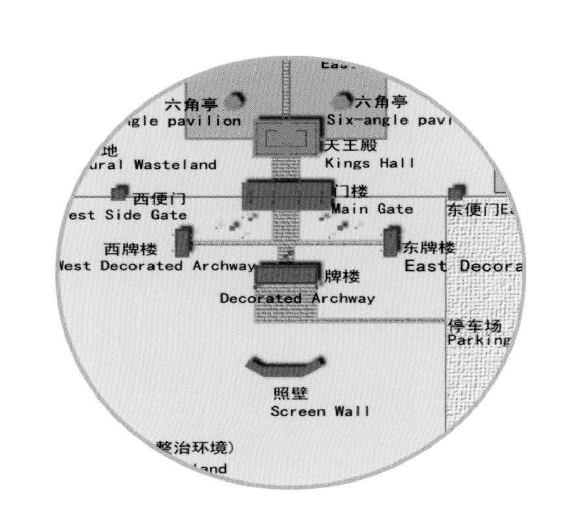

博格达汗宫博物馆平面图

Old Picture of the Front Gate

4.建筑特点

蒙元时期，统治者在蒙古高原上兴建都城，其城市布局和建筑深受汉文化影响，形成了蒙汉式建筑风格。这个时期，藏传佛教开始进入蒙古地区，开始出现寺庙。由于受西藏传统建筑和汉族宫殿式建筑风格的影响，蒙古地区藏传佛教的建筑形成了藏汉式风格，其中也融入了蒙古民族文化。

博格达汗宫博物馆建筑采用蒙藏文化风格与汉式工程做法相结合的建造手法，反映了新的民族建筑理念。主要有以下特点：

选址极佳。乌兰巴托南面之山被蒙古人尊其为“圣山”，称博格达汗山。博格达汗宫博物馆位于博格达汗山脚下的图拉河畔之临山近水、地势平坦的川地上，优雅而风景如画的环境，从地舆学角度认为是一处“风水宝地”。

总体上为典型的对称布局。博格达汗宫博物馆建筑布局深受中国的传统中轴对称制度的影响，中轴线上布列着主要建筑，两侧建筑严格对称。中路建筑为主院，左右两侧为辅院，形成“回”字形平面，成为封闭性较强的完整建筑空间，这种布局形式与藏传佛教寺院建筑的布局相似。

建筑等级制度鲜明。博格达汗宫博物馆等级最高的建筑，为二进院内的大殿，标志特征为面阔开间底层均为九间，是宫内建筑之首；它为哲布尊丹巴呼图克图礼佛、居住之所，高于其他建筑之功能；各院的功能分布、主次地位明显，如门的设置，中轴建筑的明间之门，是活佛的通行必经之门，而侧门则是供一般僧俗人的通行。

蒙汉结合的建筑构制。博格达汗宫博物馆的建筑体现以中原汉式传统建筑的基本构制为主，兼容大式、小式和民间构制于一体的一种特殊形制的建筑文化，也具有蒙古建筑文化内涵的表达。无斗拱垂柱式牌楼及牌楼门，以大式屋顶及前后垂柱形成殿式构制，下部以四柱为支撑，显得十分壮丽，这种形式属民间做法。正门、一进院、二进院中轴线上三大建筑，皆属重檐二层，但与汉式重檐二层建筑的处理手法上存在明显不同，其二层面阔间数均小于底层，宫门在顶层平顶上居中布列起建；一进院殿堂则在明间下置抱厦，上置亭式阁楼，均为一间构造；而二进主殿的二层仅为面阔三间，因施围廊做法，同时又于二层屋顶居中施以一间亭式阁楼，这种构制作法在清代官式、大式建筑中是没有的。

多元的装饰文化。博格达汗宫博物馆的内外装饰以蒙古民族特有的风格为主，同时借鉴了汉式和藏传佛教传统艺术风格的表现手法，形成了独具魅力的艺术表现风格，具有较高的艺术感染力。

建筑材料和风格反映出新的建筑理念。博格达汗宫博物馆的门前区主要建筑，由于出檐深远，梁架出挑较大，屋面选用较轻的铁皮瓦件，铁皮瓦件采用干铺的手法，无苫背层，大大减轻了屋面的荷载。在减轻屋面荷载的同时，有效地解决了传统黏土瓦件抗冻融性较差的缺点，延长了建筑寿命。这一做法充分体现出那个时代建筑师及工匠的建筑才智。

博格达汗宫博物馆的建筑，虽属批准而建，但它并非像“庆宁寺”那样，按规范图纸施建，而是采用了蒙藏文化风格与汉式工程做法相结合的手法建造，为乌兰巴托近代建筑史中一座十分完整的宫殿式建筑群。反映了清代晚期一种新的民族建筑的理念，保存至今近120年而基本完好，体现和代表了当时建筑科学技术的水平，具有一定的科学价值，在蒙古地区建筑史上占有一定地位而具有一定的研究价值。

5.彩画的艺术特点

中国建筑彩画是中国古代建筑特有的装饰形式，有极强的等级观念和丰富多彩的图案，是中国古代建筑的重要组成部分。古代建筑上的彩画主要绘于梁和枋、柱头、窗棂、门扇、雀替、斗拱、墙壁、天花、瓜筒、角梁、椽子、栏杆等建筑木构件上。主要以梁枋部位为主。

建筑彩画最早可以追溯到西周时期，经秦、汉、魏、晋、南北朝、隋、唐、宋、元、明、清等朝代，逐渐由简单到复杂，由低级到高级。

秦汉时期在宫殿的柱子上涂丹色，在斗拱、梁架、天花等处施以彩画，装饰图案多用龙、云纹，并且逐渐采用了“凌锦”织物纹样，称为锦纹。

在六朝和隋唐时期彩画开始向着成熟期发展了，纹饰上已形成一种像是花边似的，二方连续、四方连续两种纹饰。受佛教艺术的影响，又产生了新的建筑装饰图案，如卷草纹、火焰纹、莲花、宝珠、曲水和卍字等等。这些图案影响深远，其中云纹、龙纹、凤纹、凌锦纹、卷草纹和卍字纹始终是彩画中的主要和典型图案。

宋元是中国彩画成熟时期，构图基本成熟了，檩、梁、枋、柱有四种形式,斗拱是重点装饰部位，上面遍绘纹式、花。支条、天花也是重点装饰部位，中间花饰成型了。从色彩上说：冷暖色调并存，为后来的彩画以冷调子为主奠定了基础。这些在宋代《营造法式》中可见到。元代又出现了旋子彩画，但还不成熟，整体色调已经由早期多用暖色转以青绿等冷色为主。

到了明、清阶段，彩画发展到了它的鼎盛时期，在继承传统的基础上，取材和制作方面又有了新的变化与发展，集历代彩画之精华，新的品种不断涌现，题材不断扩大，表现手法不断丰富，法式规矩也更加严密规范。

明代梁枋彩画的格局继承了元代，并有了不小的发展，保留了构图三段式。色彩特征极其明显，转向以蓝绿为主的冷色调，色彩淡雅。到清代时期，尤其清代晚期，由自由化逐步走向了程式化，建筑彩画名目繁多，一般分为三大类，即旋子彩画、和玺彩画及苏式彩画。坛庙、寺院多用旋子彩画，宫殿建筑多用和玺彩画，苏式彩画则用于一般园林住宅建筑。和玺彩画是古代彩画中规格最高的彩画，用于宫殿中。枋心绘行龙或龙凤图案。根据不同内容，和玺彩画分为“金龙和玺”、“龙凤和玺”、“龙草和玺”等不同种类。

旋子彩画，又称学子、蜈蚣圈，是中国古代建筑上古建彩画风格的一种，在等级上仅次于和玺彩画，可广泛见于宫廷、公卿府邸。旋子彩画因藻头绘有旋花图案而得名。

苏式彩画，又称苏式彩绘，是中国古代建筑彩画的一种，发源于南方。等级低于前两种。画面为山水、人物故事、花鸟鱼虫等。

在博格达汗宫博物馆可以看到和玺彩画、旋子彩画和苏式彩画三种形式，但其主要以延承清代晚期的苏式彩画为主。苏式彩画主要采用了枋心式、包袱式彩画的构图形式，内容包括人物、花鸟、山水，题材取自戏剧、小说以及现实的写实绘画，绘画技法多样，包括烘染、晕染、工笔、写意、兼工代写及白描勾线后进行局部单色的罩染，有一部分为洋抹，即吸收了国外的绘画技法而形成的一种画法，主要体现在一些风景画中 ，画面用颜色以青绿二色为主色，根据画面的需要使用了其他大量的配色，并且使用了大量的沥粉贴金工艺。其中，博格达汗宫博物馆主楼三扇大门分别绘有左右门神共计大小六尊，构图铺满整个平面，背景单纯，颜色运用沉稳中又有对比，显示出威仪的神态。

博格达汗宫博物馆古建筑彩画整体图案造型简洁，构图设色富有变化，图案内容丰富，色调清雅活泼，装饰效果严谨富丽。大量的沥粉贴金增强了视觉的效果。

2.Historical background and architectural art

Bogd Khaan Palace Museum was first built in 1893, the eighth Bogd Jivzundamba Hutukutu, or Bogd Khaan, the highest religious leader of Mongolia, once lived there and conducted religious activities. In May 1924, the eighth Bogd Jivzundamba died. Bogd Khaan Palace was converted into a State Museum in April 1st 1926 preserving many of the living objects owned or used by the Bogd Khaan, religious objects, traditional drawings and some precious cultural heritage from 17th to 20th centuries.

1. Historical Record

The eighth Bogd Jivzundamba was born in Lhasa, Tibet in 1869. In 1871, his identity as the eighth reincarnation of the Bogd Jivzundamba was confirmed. In October 1874, the Bogd Jivzundamba left Tibet with his parents and siblings to take residence in Ulaanbaatar. Upon his arrival, he was greeted with considerable welcome from aristocrats and worshippers from across the country. Then he became the religious leader of Mongolia. Following the collapse of Manchu empire in 1911, he obtained political authority over Mongolia as Bogd Khaan, he died in 1924.

Bogd Khaan Palace Museum experienced three different political revolution stages from later period of Qing Dynasty to People' s Republic and Mongolia independence until the eighth Bogd Jivzundamba' s death in 1924. Bogd Khaan Palace Museum witnessed the history from submitting to Qing Dynasty after Mobei time to establishing an independent state. As a direct historical witness, Bogd Khaan Palace Museum, which bearing considerable historical value and research value, had vital function for history supplementation , authentication and correction.

2.Geographic location

Bogd Khaan Palace Museum is located at the flat land near Tula riverside at the foot of Mountain Bogd Khaan in the south outskirts of Ulaanbaatar (the capital of Mongolia) with picturesque surroundings.

3.Architectrual Structure

The plane of Bogd Khaan Palace Museum shows a layout of font back and rectangle. Three groups in cross direction, the summer palace is in the middle where once lived Bogd Jivzundamba Hutukutu. The right and left groups locates at east and west side of the middle. Outside the palace gate are the square and facilities such as screen walls and wooden archways. Wooden fence wall can be also found at the square.

The architectures in the middle are successively the square, the gate, ZhongGu pavilion, the first courtyard, the second courtyard, the third courtyard. The second courtyard is the main part of the palace where Bogd Jivzundamba Hutukutu once lived, which is also called summer palace.

The timber bounding wall is outside the left and right courtyard of Bogd Khaan Palace Museum. There are two side doors respectively sit at the two sides in the middle of timber bounding wall. The left one is a little bit to the south including a two-floor winter palace with Tibetan style which face the west, this palace is originally two-floor Tibetan style, after rebuilding, the gate and window are shifted into

Russian style, the flattop is also changed into simple Xieshan style.

4. Architectural features

During Mongol-Yuan period, the rulers built capital on the Mongolian Plateau, the urban layout and architectures were mostly influenced by Han culture, which led to the formation of the Mongolian and Chinese architectural styles. At this period, Tibetan Buddhism started to enter Mongolia region and the temple appeared. Due to the influence of traditional Tibetan architectures and the Han palace, the buildings of Tibetan Buddhism in Mongolia formed the Tibetan and Chinese style, the Mongolian culture was also integrated.

Bogd Khaan Palace Museum is a combination of construction techniques of Mongolian-Tibetan cultural style and Han engineering practices, which reflects the new national architectural concept. The main characteristics are as follows:

Excellent location. The mountain of the south of Ulaanbaatar is respected as the "Holy Mountain" by Mongolian. Bogd Khaan Palace Museum is located at the flat plain near Tula riverside at the foot of Mountain Bogd Khaan in the south outskirts of Ulaanbaatar with picturesque surroundings, which is regarded as "treasure place".

The typical symmetrical layout. The layout of Bogd Khaan Palace Museum is deeply affected by Chinese traditional axial symmetry system, the main buildings sit at the central axis with strict symmetry of both sides of the buildings. The middle buildings are chief parts, while the left and right side are supplemental parts, which forms the plane of square shape with strong closure, this form of layout is similar to the layout of Tibetan Buddhism temple buildings.

Distinct architectural hierarchy. The highest rank of the construction in Bogd Khaan Palace Museum is the main hall in the second courtyard with marking features where Bogd Jivzundamba Hutukutu lived and conducted religious activities , it is the first class among all the buildings; the functional distribution of each courtyard is obvious, such as the set of doors and the gate at central axis, it is the passage which living Buddha pass through, while the side doors are for monks and lay people.

A combination of Mongolia and Han architectural structure. Buildings of Bogd Khaan Palace Museum reflect the basic configuration of traditional Han architectural culture, which is compatible with big-style, small-style and folk structure, and also an expression of cultural connotation of Mongolia architectures. Vertical column archway without brackets and archway doors are built in the structure of big-style roof and the front-rear vertical column with four pillars for support below which appears to be very spectacular, this form is a folk practice. The front gate, the first courtyard and the three buildings at central axis of the second courtyard are all double eaves and two-story style, but there are obviously differences from double eaves and two-story buildings of the Han style on the practice, the number of each building width of the second story is less than the bottom; baoxia is built below the central bay of the hall at the first courtyard, and above it is pavilion-style attic; while the second story of the main hall at the second courtyard is only three building width, which can not be seen in official-type and big-type building of Qing Dynasty.

Diverse decorative culture. Both interior and exterior decorations of Bogd Khaan Palace Museum show the unique style of Mongolian, drawing on the performance practices of Chinese type and Tibetan Buddhist tradition, which form charming artistic expression with high artistic infection.

Building materials and style reflect the new architectural concept. Because of the far-reaching eaves and beams, lighter iron sheet tiles are used by way of dry-paved without straw mat layer on the roof of the main buildings in front of the Bogd Khaan Palace Museum. This method reduces the roof load and effectively solves the problem of traditional poor freeze-thaw resistance, and extends the life of the buildings. This approach fully reflects the architectural talents of architects and craftsmen of that era.

Although got the approval, the buildings of Bogd Khaan Palace Museum are not like the "Qing Ning Temple" which is constructed according to the specification drawings, instead, they are a combination of techniques of Mongolian-Tibetan cultural style and Han engineering practices, which are a group of very integrated palace-type architectural complex in modern history in Ulaanbaatar. These buildings, preserved nearly 120 years and largely intact, reflect the architectural philosophy of the late Qing Dynasty and represent the level of building science and technology of that time. Moreover, the buildings of Bogd Khaan Palace Museum occupy a certain position in the building history of Mongolia region and possess a certain scientific and research value.

5. The artistic characteristics of colored drawings

Chinese construction colored drawing is a unique decorative form of Chinese ancient architectures which reflects a strong sense of hierarchy and possesses a variety of patterns, it is an important part of Chinese ancient architectures. The paintings of ancient architectures are mainly painted on roof beam, square-column, column head, window lattice, door leaf, QT, bracket, wall, ceiling, melons tube, corner beam, rafter, railing and other wood components. Roof beam and square-column are the main parts.

The earliest architectural colored drawings can be traced back to the Western Zhou Dynasty and undergo the Qin, Han, Wei, Jin, Southern and Northern Dynasties, Sui, Tang, Song, Yuan, Ming and Qing Dynasties, the drawings develop gradually from simple to complex, from lower level to higher level.

In Qin and Han Dynasties, Dan color (Chinese traditional color, look like the color of sunrise) was painted on the pillars of the palace and colored drawings on the brackets, beams and ceilings etc., the decorative patterns were almost dragon and cloud with gradual introduction of "Ling Jin" fabric patterns which called Jin pattern.

The development towards maturity of colored drawings took place in Six Dynasty and the Sui and Tang Dynasties, the ornamentation on the formation was developed a kind of lace-like pattern. New construction decorative patterns appeared due to the effect of Buddhist art, such as curly grass pattern, flame pattern, lotus, Qushui and so on. These patterns had a far-reaching implications, dragon, phoenix, Lingjin pattern and curly grass pattern were always the main and typical patterns in colored drawings.

Song and Yuan Dynasties are the mature period for Chinese colored drawings, the composition is basically fine, purlins, beams, square-columns, and pillars have three forms, brackets are the key decorative parts, on them are painted decorations and flowers. Branch and ceilings are also the key parts for decoration. The coexistence of warm and cold colors lay the foundation for the later paintings of the cold tone. These can be seen in <Ying Zao Fa Shi> of Song Dynasty. Xuanzi colored paintings appear in Yuan Dynasty, but not yet mature, the overall color tone is more warm color and then turn to cold color such as blue green.

To the period of Ming and Qing Dynasties, the colored drawings develop to great prosperity, on the basis of the tradition, material picking and picture producing undergo a new change and development, gathering the essence of the colored drawings of past dynasties, the new varieties are emerging and the themes are constantly expanding, also are the expression art.

Colored drawings on beams in Ming Dynasty inherit that of Yuan Dynasty, but develop and retain the three stages of composition. The color features are obvious, turn to cold color of blue and green which look elegant. When it goes to Qing Dynasty, especially late Qing, the liberalization is gradually moving towards stylization, the type of the construction drawings is various, which can be generally divided into three categories as Xuanzi colored drawing, Hexi colored drawing and Su style colored drawing. Xuanzi colored drawing is often used in temples while Hexi colored drawing largely for palatial architectures, and Su style colored drawing generally for garden residential construction.

Hexi colored drawing is the highest level in ancient paintings and mainly for palace. The patterns are largely dragon and phoenix. Hexi colored drawing is divided into different types according to different contents.

Xuanzi colored drawing, also known as Xuezi or centipede circle, is a kind of ancient colored painting style on ancient Chinese architectures and rank the second to Hexi type, Xuanzi type can be widely seen in royal palace and mansions. Xuanzi colored drawing gets its name for xuan flower painted on the beam.

Su style colored drawing is a kind of Chinese ancient architectural pattern, it originates from the south and ranks lower than the two types above. The contents are landscape, stories of people, flowers, birds, fishes and insects, etc.

Hexi colored drawing, Xuanzi colored drawing and Su style colored drawing can be seen in Bogd Khaan Palace Museum, but they mainly inherit Su style in late Qing. Su style colored drawing chiefly adopts the composition form of fangxin type and baofu type with patterns of people, flowers, birds and landscapes, the themes are from dramas, fictions and realistic paintings with various painting techniques such as shading, blending, elaborate style, liberal style, freehand brushwork and line drawing, partly is yangmo, which absorbs the painting techniques of the foreign and mainly reflects in some of the landscapes pictures with blue and green and other colors according to the need of the screen, the craft of gold powder paste is also introduced. There are six door-gods on the three door leafs of the main building of Bogd Khaan Palace Museum, the composition covers the entire plane with simple background, the colors used are both clam and contrastive, showing the demeanor of dignity.

The overall pattern of colored drawings on ancient buildings of Bogd Khaan Palace Museum is simple shaped, the composition and color are full of change, the pattern is rich in content, color tone is elegant and lively, decorative effect is rigorous and rich. A large quantity of gold-powder paste enhances the visual effects.

大门老照片
Old Picture of the Front Gate

中牌楼老照片
Old Picture of Middle Archway

博格达汗宫博物馆位于“圣山”——博格达汗山脚下
The Location of Bogd Khaan Palace Museum: at the foot of Mountain Bogd Khaan or called “The Holy Mountain”

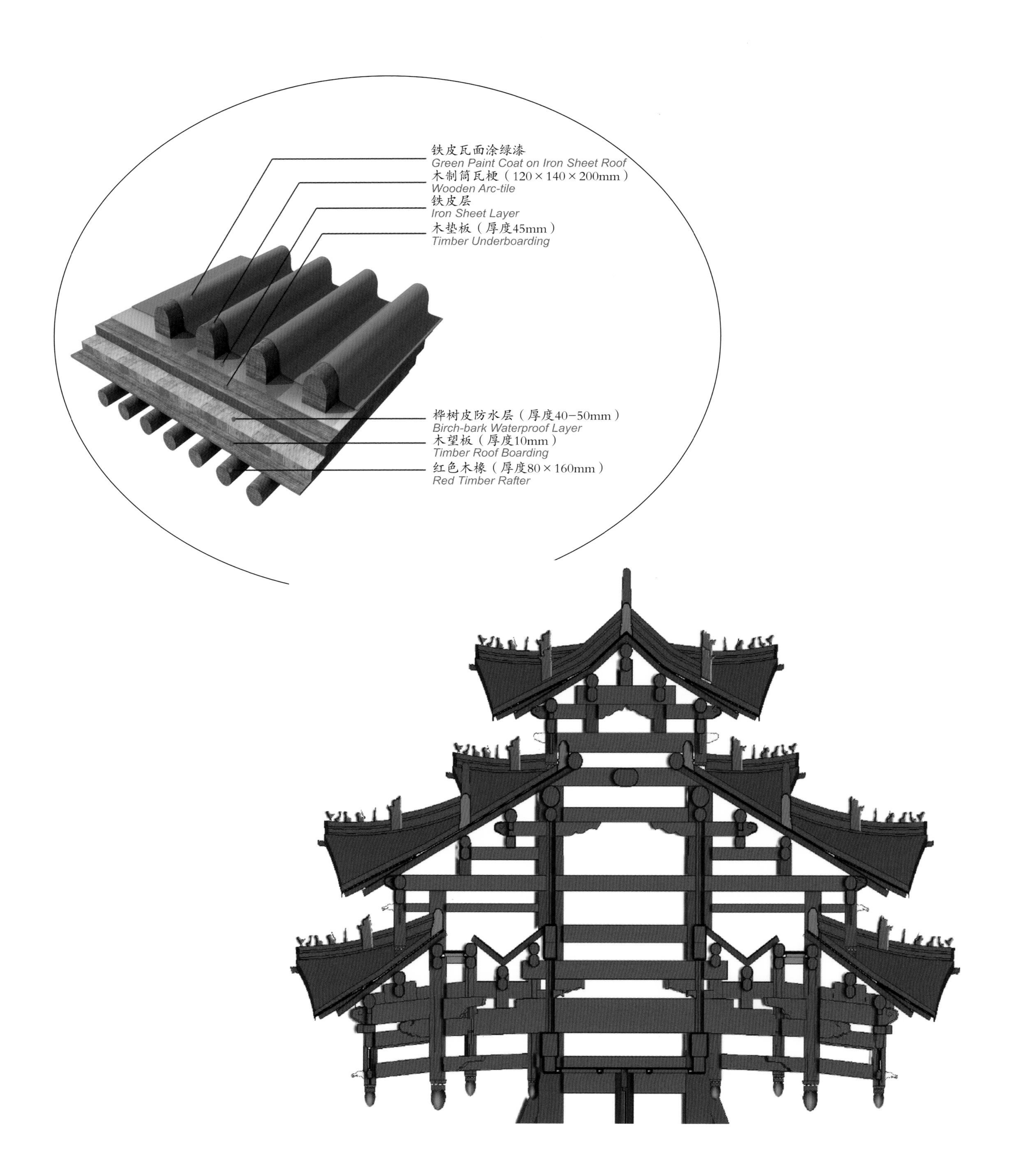

博格达汗宫博物馆的建筑材料和风格

Building Material and Architectural Style of Bogd Khaan Palace Museum

该建筑群具有藏传佛教建筑的特色
The Architectural Complex possesses features of Tibetan Buddhism Architectures

藏传吉祥八宝—白海螺图
One of The Eight Auspicious Symbols from Tibetan Buddhism alled "White Conch Picture"

苏式彩画花鸟图
Painting of Flowers and Birds in traditional Su style

沥粉贴金金龙——等级象征
Gold-plated Dragon(the symbol of the social hierarchy)

童趣图
Children Playing Picture

吸毒疗伤图——古代故事
Drug Healing Picture (a tale of the past)

苏式彩画花草图
Painting of Flowers in traditional Su style

西游记故事图、山水图
Picture of Pilgrimage to the West, Picture of Landscape

叁 Three

工程维修

PROJECT IMPLEMENTATION

工程维修

PROJECT IMPLEMENTATION

The Maintenance Project of Bogd Khaan Palace Museum

博格达汗宫博物馆维修工程

3 工程维修

博格达汗宫博物馆以大门及南宫墙为界，北为宫殿区，南为门前广场区。本次项目维修的范围为门前广场区。门前广场区占地面积8600平方米，广场中心有照壁一座，向北依次为牌楼一座，大门一座，大门与中牌楼之间的东西道路两端相对各置小牌楼一座。大门东、西两侧宫墙各设便门一座。大门前东、西两侧设旗杆两根。本次维修项目分为古建筑本体维修和油漆彩画维修两大部分。

1.工程时间

2005年5月，设计人员赴现场进行勘察；

2005年10月，完成门前区测绘图及初步维修方案的设计；

2005年11月，维修方案通过了由中国国家文物局组织的专家评审；

2006年5月，保护维修施工图正式提交蒙古国相关专家并审核通过；

2006年5月工程正式开工，2007年9月完工。

2.本体维修

博格达汗宫博物馆门前区古建筑由于建造年代久远，存在着大量隐患和病害。包括：大木结构老化、糟朽、劈裂、变形；屋面变形渗漏、椽子望板糟朽、木架歪闪；建筑基础沉降、墙体开裂坍塌；院落排水不畅；砖、石严重风化、彩画老化褪色等病害。对古建筑的保护维修的目的，就是要利用科学的方法来保护古建筑，使之能“益寿延年”。维修的原则就是要不改变原状，“原形制、原结构、原材料、原工艺”，是维修工程的四大施工原则。博格达汗宫博物馆门前区古建筑维修，属于全面整治工程。即对古建筑主体、附属设施及周边环境进行的以恢复古建筑整体风貌、合理使用为目的的全面的、保护性修缮整治工程。

2.1 前期勘察

在前期勘察中，中国国家文物局和西安文物保护修复中心派文物保护专家赴蒙古国对博格达汗宫门前区古建筑进行了现场实测。经现场实测，大门、牌楼、宫墙及东西便门等木构建筑大木构架基本完好，主楼梁架部分变形下垂；因抱头梁、穿插枋年久变形下沉，导致垂莲柱倾斜，花板局部破损；个别木柱柱身开裂、柱础下沉，原夹杆石被改为混凝土夹杆石；屋面铁皮瓦面锈蚀、脱漆严重；木板墙体木质糟朽、倾斜严重；条砖地面凹凸不平，水泥阶沿风化散裂；木装修上、下槛倾斜，导致大门走闪，开启困难；大木下架油漆地仗起皮开裂；彩画画面褪色、龟裂严重；地面凹凸不平，无台明，无散水。照壁一字部分因基础不均匀沉降，导致两端须弥座、墙体、墀头、瓦顶、正脊开裂。裂缝内因雨水侵入，冬季冻蚀膨胀，导致西侧墀头向南倾斜；须弥座下北面青砖土衬有不同程度的硝碱残破，南面下碱部分因硝碱、开裂，现用水泥砂浆罩面；四周方砖散水有不同程度的硝碱残破。

3.Project implementation

The front gate and south palace wall are the boundary of Bogd Khaan Palace Museum, the north area is palace district and the south is square region. The maintenance area is the square in front of the gate. The front square covers 8600 square meters, the screen wall sits in the middle of the square, to the north are decorated archway and front gate, tiny archways sit at both sides of the east-west path which is in the middle between the front gate and middle archway. There are side doors at both east and west palace walls, two flagpoles can be seen at east and west sides in the front gate. The maintenance projects contains architecture maintenance and colored drawings maintenance

1. Project Time.

In May 2005, designers conducted field investigation; In Oct. 2005, the front square surveying and mapping diagram and preliminary maintaining design was completed; In Nov. 2005, the maintaining plan was passed by experts review organized by State Administration of Cultural Heritage; In May 2006, maintaining chart was officially sent to Mongolian experts and it got passed finally. The maintenance project successfully came into operation in May 2006 and was perfectly completed in Sep. 2007.

2. Maintenance Project

Due to the long period establishment of the ancient architectures in front of the Bogd Khaan Palace Museum, a great quantity of hidden problems and diseases still exist which include: wooden component aged, decayed, cracked and deformed; the roofing deformed and leaked, rafter and roof boarding decayed, timber frame slanted; the construction foundation sunk, the wall body cracked and fell down; the courtyard drainage is poor; bricks and stones severely air-slacked, colored drawings aged and faded, etc. The aim of ancient architectures protection and maintenance is to use scientific approaches to protect them, giving them a prolong life. The principles of the maintenance is to refurbish without changing their former looks, “ former shape, former structure, former materials, fomer crafts ” are four construction principles. The maintenance of the ancient architectures in front of the Bogd Khaan Palace Museum is a comprehensive renovation project, which is to recover the general styles and features of the ancient architectures , subsidiary facilities and the surroundings.

2.1 Preliminary survey

Cultural heritage protection experts, assigned by Chinese State Administration of Cultural Heritage and Xi' an Center for the Conservation and Restoration of Cultural Heritage, conducted field survey on ancient architectures at the front square of Bogd Khaan Palace Museum. The result indicates that wooden structure of the front gate, archways, palace walls and east-west side doors were basically fine, beam frames on main building partly deformed and sagged; due to the distortion and subsidence of baotou beam and penetrating tie, lotus column leans and the board were damaged; some timber columns cracked, the column foundation sank, former pole clamping plinth were replaced by concrete ones; iron sheet tiles on the roof were corroded, the paint coat dropped severely; wooden board walls were decayed and leaned severely; brick floor was uneven, the edge of cement platform was air-slaked and destroyed; headsill and soleplate of wooden decoration leaned which made the front gate distorted and opened uneasily; the base layer of the painting on bottom-beam peeled and cracked; colored drawings faded and cracked severely; the ground was uneven without platform and drainage platform. The screen wall sank because of asymmetrical base which results in the crack of pedestal of Buddha statue, wall body, eave tile, tile peak and main ridge. Being eroded by rainfall, the fissures freeze and expanded in winter which caused west eave tile leaned 8.5cm to the north; blue bricks of north side under pedestal of Buddha statue appeared varying degrees of damage, south of them cracked and had been covered by cement plaster; quadrel and drainage platform around all showed different degrees of damage.

门前区
The front square

大门

大门位于博格达汗宫中轴线上门前区偏北位置，坐北面南，面阔三间，通面阔16.6米，进深7.2米，高11.8米，建筑面积173.47平方米。大门主楼为重檐五滴水歇山顶屋面，次楼为重檐二滴水歇山顶屋面，设圆椽、飞椽及望板，无苫背层，干铺铁皮砸制成的筒板瓦形成瓦件，以绿漆刷饰。上设的脊兽、小五兽、山花等均为铁皮砸制而成。

大门1912年至1919年间修建，亦称安定门、和平门。有着独一无二的特征，即没有使用一枚钉子，用108种不同规格的构件连接而成，装饰具有博格达汗时期蒙古艺术的典型特征。

The front gate

The front gate locates in the central axis, and little north to the entrance of Bogd Khaan Palace Museum, locates in north and faces to south with three kind of spacing. The axial distance between end pillars is 16.60 meters, deep into the 7.2 meters, 11.8 meters high, with a construction area of 173.47 square meters. The main archway of the gate is in double eave roof, five drip tile, gable and hip roof. The subordinate archway is in double eave roof, two drip tile, gable and hip roof. They contain round rafter, flying rafter and eave board, but without thatch layer. The tile is made up of round tile which compressed by iron sheet, and painted green. The special animal and five kinds of animal, and pediment, etc are all manufactured by iron sheet.

This gate was constructed between 1912 and 1919, also called Andi Men, or Peace Gate, The gate has the unique characteristic of being constructed without a single nail, using 108 different forms of interlocking joints instead. Decorations on the Peace Gate showed typical features of Mongolian art in the period of Bogd Khaan.

大门
The Front Gate

东、西便门

东、西便门位于博格达汗宫大门东、西两侧，距大门为29.7米。东、西便门为悬山顶屋面，设圆椽、飞椽及木望板，无苫背层，用铁皮砸制成的筒板瓦形式瓦件，以绿漆刷饰。上设的脊兽等均为铁皮砸制而成。面阔2.23米，进深2.95米，高4.42米。柱外两侧设木装板墙，墙顶为板檐。

East and west Side Doors

East-west side doors are located at the east and west side of the front gate of Bogd Khaan Palace Museum which are 29.70 m away from the gate. The roof on east-west side doors is overhanging gable type with round rafter, flying rafter and roof boarding on it, the roof is paved by arc-tiles painted with green color, the ridge decorations are wrapped by iron sheet.,2.23 m width and 1.45 m depth, eave board wall is at the two sides of column with board eave to be as the wall top.

东便门
East Side Door

西便门
West Side Door

中、东、西牌楼

牌楼位于博格达汗宫大门与照壁之间，共三座。中牌楼位于中轴线上，东、西牌楼位于中牌楼偏北两侧，对称设置，相距51.7米。中牌楼面阔10.32米，进深4米，高8.9米，建筑面积50.88平方米，立面为明次楼三间加斜杆、悬挑式构造。东西牌楼轴线宽度为3.75米，进深4.14米，高5.92米，建筑面积45.6平方米，立面为两侧加斜杆支撑的古建大式做法。东西牌楼及中牌楼屋面形式均为歇山顶，采用铁皮砸制成筒板瓦形，刷饰绿漆，勾头、滴水、脊兽及小五兽、博风排山沟滴等均为铁皮制成。

Three Decorated Archways at middle, east and west

Decorated archways sit between the gate of Bogd Khaan Palace Museum and the entrance screen wall, three archways in total. The middle archway sits in the central axis, east and west ones sit symmetrically and little north at the both sides of the middle one, with the distance of 51.7m. The elevation of the middle archway is built up with three subordinate archways and slanting pole in cantilever pattern, of 10.32 m width, 4 m depth and 8.9 m height with construction area of 50.88 square meters. The axis width of east and west archways is 3.75m, 4.14m depth and 5.92m height with construction area of 45.6 square meters. The elevation of the east and west archways are constructed with slanting poles by two sides in wooden frame of ancient architectural pattern. The roofing pattern of the east, west archways and middle archway are gable and hip roof. They take the iron sheet into barrel flat tile and then paint green. The eave tile, drip tile, special animal and five kinds of animals, gable eave board, paishan, eave and drip tile, etc are all manufactured by iron sheet.

中牌楼
Middle Archway

东牌楼
East Archway

西牌楼
West Archway

旗杆

立于大门两侧，高18米，是用来悬挂蒙古国旗和宗教旗帜的标志性设施，西侧的旗杆悬挂着博格达汗时期的蓝色蒙古国旗，东侧旗杆上则飘扬着黄色的宗教旗帜。体现了蒙古在博格达汗时期是属于政教合一的体制。

Flagpole

The Flagpoles stand symmetrically at both sides of the front gate, which are used to hang up the Mongolian national flag and religious flag, on the west pole, the blue national flag of Mongolia under the Bogd Khaan’s rule was hung up there, whereas a yellow religious flag was shown on the east pole, indicating the political system of the unification of between the state and the religion at the period of Bogd Khaan’s ruling.

旗杆
Flagpole

照壁

照壁位于博格达汗宫中轴线上门前区南段，形式为青砖八字照壁。一字部分面阔9.1米，壁厚0.93米，高7.06米。八字部分每边面阔3.18米，壁厚0.67米，高5.03米，通面阔16.14米。照壁下部为青砖浅雕须弥座，上、下枭浅雕巴达马图案，束腰部分采用竹节柱子间隔，柱间作夔龙图案。上部北面照壁心内作浅浮雕山、祥云、海水及二龙戏珠图案，南面为方砖心墙。边框雕蕃草图案，照壁心框外两边作柱子，柱外砌撞头。边框以上起枋子、垂柱，枋下饰流苏、团花图案。上身至瓦顶之间下作三踩如意斗拱，斗拱上作冰盘檐椽子、飞椽。八字部分上身，正、背面均为方砖心墙，边框外砌撞头，其上起枋子、垂柱。屋面前后檐均为硬山式，两侧山花博风砖，筒板瓦灰陶屋面，脊兽为五脊之兽，为我国清代建筑形制。

Entrance Screen wall

The entrance screen wall locates at south end of central axis at front square of Bogd Khaan Palace Museum. The shape looks like Chinese character "eight". Linear shape section is 9.1 m width, 0.93 m thick and 7.06 m height. "Eight" shape part is 3.18 m width at each side, 0.67 thick, 5.03 height and 16.14 m in total width. The bottom of the screen wall, which is blue brick shallow carving pedestal of Buddha statue with badama design and dragon pattern on it. The central of north screen wall is carved with mountain, auspicious clouds, seawater and pattern of two dragons playing with a pearl, the south side is quadrel core-wall. Grass pattern is carved on the border, the two sides of frame of central screen wall are used for column with eave tiles outside. Tie-beam and column can be seen above the frame, below the tie-beam is decorated by tassels and flowers. There appears Ruyi bracket, roof rafter and flying-rafter. The upper body, the right and back sides of "Eight" shape partare quadrel core-wall, eave tiles are built outside the border with tie-beam and column above it. The appearance style of screen wall roof is yingshan type with pedimental gable eave bricks and arc-tile grey pottery roof, ridge decoration are beasts, which is the architectural characteristic of Qing Dynasty.

照壁
Entrance Screen wall

古建筑主要残损表

		名称	建筑面积（m²）	地面、墙体	屋面	木构架	木装修	油漆彩画	备注
1	院前—1	大门及两侧门楼	173.47	土地面坑洼不平，门楼周围杂草丛生	铁皮瓦部分生锈约占屋面30%	1.外大门处柱子下沉，导致柱头上额枋弯曲变形；2.垂花柱头已残缺	1.部分装修板有残损；2.博风板残破严重	梁枋上彩绘褪色、柱上油漆老化部分剥皮脱落	
2	院前—2	中牌楼	50.88	1. 地面砖坑洼不平，牌楼周围杂草丛生；2. 后加水泥墩（原来没有）	基本完好	梁枋上彩绘褪色、木柱油漆老化部分剥皮脱落	总体完好，木雕刻花有部分残损	梁枋上彩绘褪色、柱上油漆老化部分剥皮脱落	
3	院前—3	牌楼东、西	45.60	1. 青砖铺地残破不平（席纹）；2. 后加水泥墩（原来没有）	铁皮瓦部分生锈约占屋面30%	完好	完好	梁枋上彩绘褪色、柱上油漆老化部分剥皮脱落	
4	院前—4	木板墙	85M（31间）	土地面坑洼不平，杂草丛生	木屋面干裂变形	木柱腐朽干裂变形	木板变形残破严重	柱梁枋上油漆老化部分剥皮脱落	
5	院前—5	照壁	21.63	1. 地面砖坑洼不平，照壁周围潮湿，杂草丛生；2. 墙体裂缝多处，砖刻有残损、墙体有倾斜，倾斜10度，墙体扭闪	照壁屋面为灰陶瓦，有部分瓦件、脊兽残缺	无	无	无	该建筑近几年曾做维修

Table of damaged parts of the front square of Bogd Khaan Palace Museum

		Location	Area (m²)	Ground、Wall	Roof covering	Wooden structure	Timber decoration	Colored painting	Note
1	Front square-1	Gate and archways at two sides	173.47	Ground uneven, weeds around archways	Iron sheet tiles rust 30% of roof	1.Column outside gate sagged, cause architrave deformed; 2.Chapiter damaged	1.Some decorate board damaged; 2.Gable eave board severely damaged	Colored painting on beam fade, aged painting on column peel and drop	
2	Front square-2	Middle archway	50.88	1.Foor tile uneven, weeds around archways; 2.New cement pier (it used not)	Basically fine	Colored painting on beam fade, aged painting on column peel and drop	Totally fine, Some wood carving design damaged	Colored painting on beam fade, aged painting on column peel and drop	
3	Front square-3	East-west archway	45.60	1.Blue brick ground broken and uneven(basket); 2.New cement pier (it used not)	Iron sheet tiles rust 30% of roof	Fine	Fine	Colored painting on beam fade, aged painting on column peel and drop	
4	Front square-4	Timber wall	85M (31)	Soil ground uneven, weeds everywhere	Timber roof suncrack and deformed	Wooden column decayed, suncrack and deformed	Board severely damaged and deformed	aged painting on beam peel and drop	
5	Front square-5	Screen wall	21.63	1.Ground uneven, moist and weeds around screen wall; 2.Several fissures, brick design damaged, wall deformed and leaned 10 degrees	Screen wall roof covered by terracotta, some tiles and ridge decorations damaged.	None	None	None	This building being repaired rent year

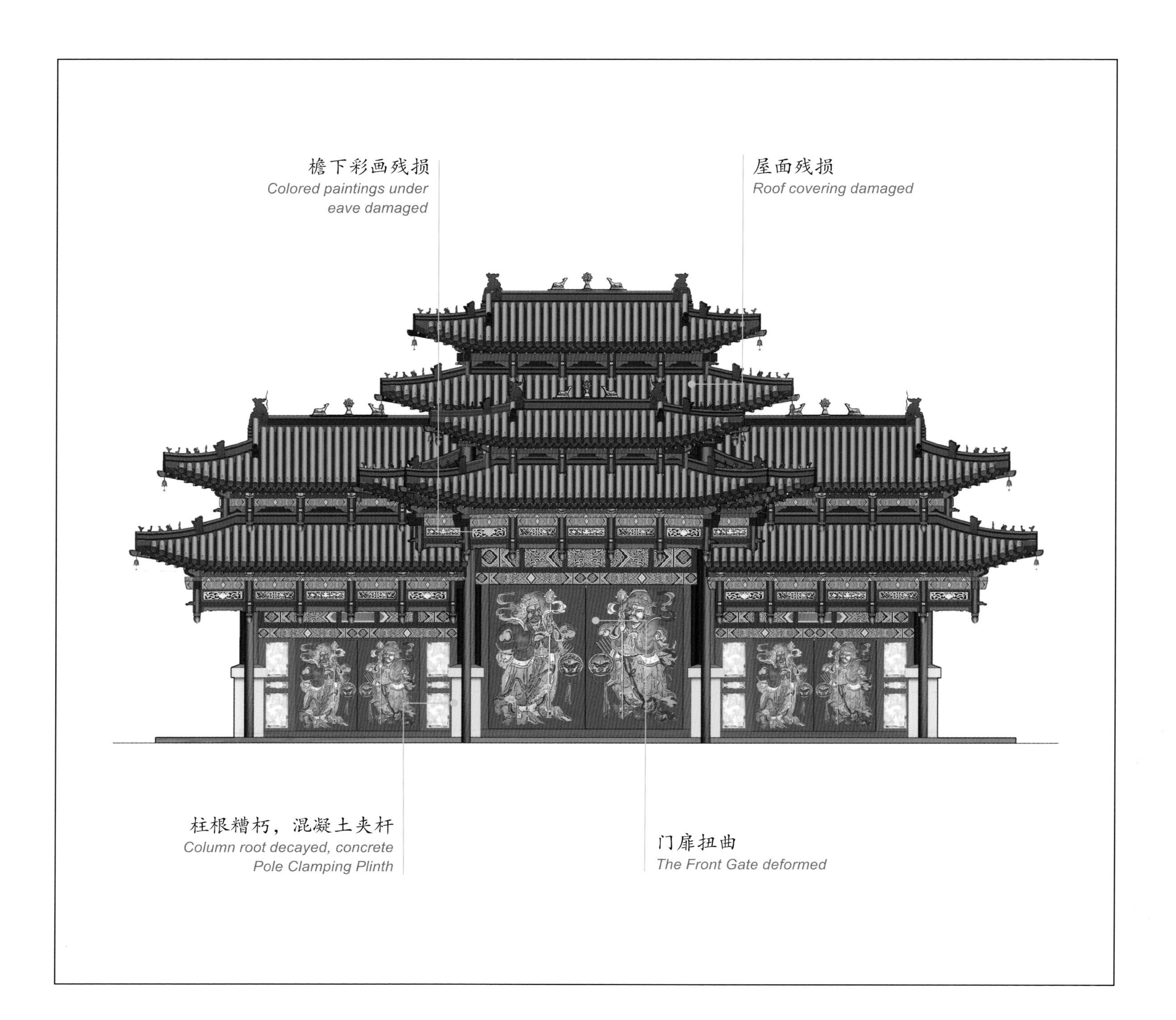

大门正立面
Front Vertical Face of the Front Gate

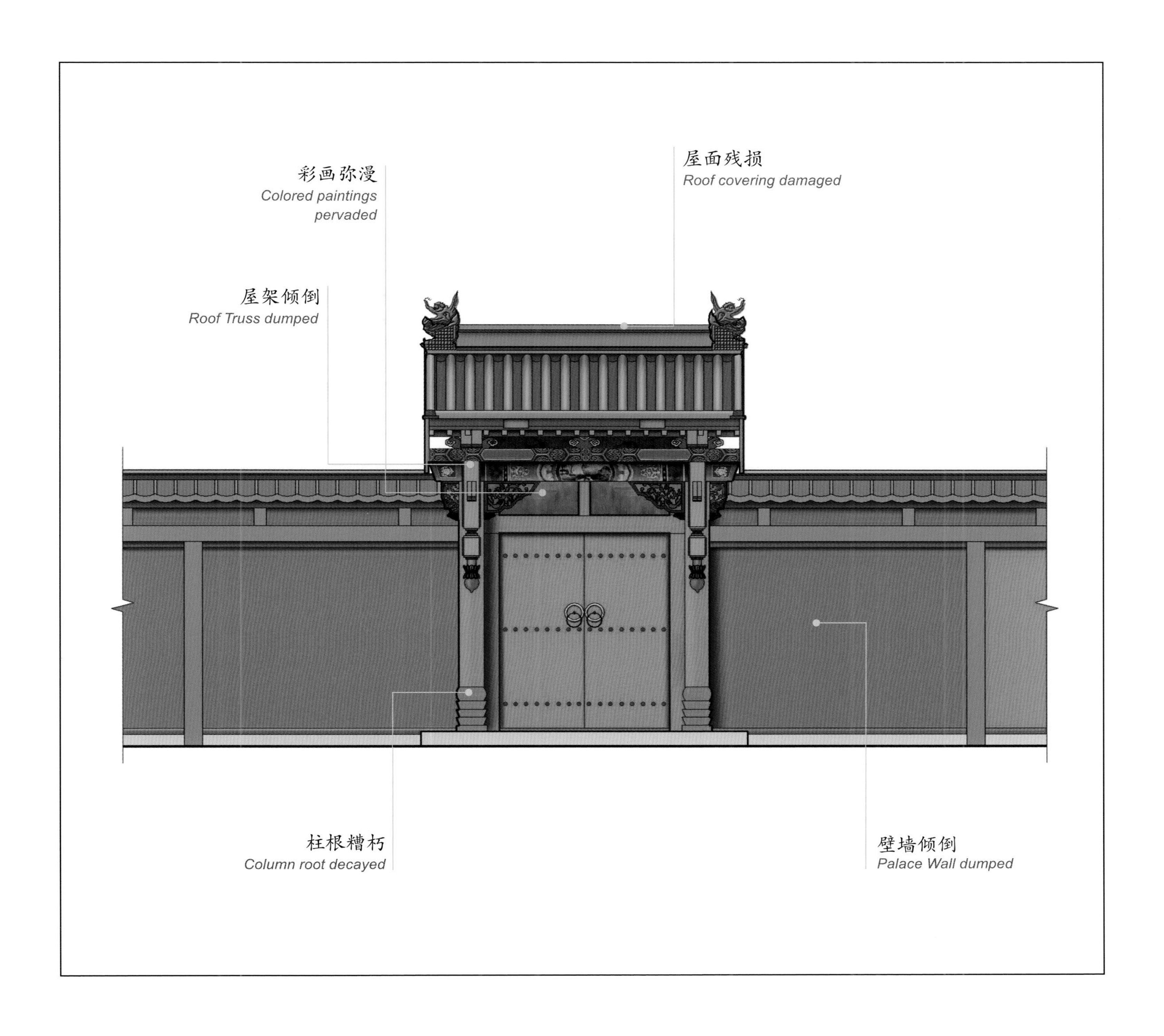

东西便门的正立面
Front Vertical Face of east-west Side Doors

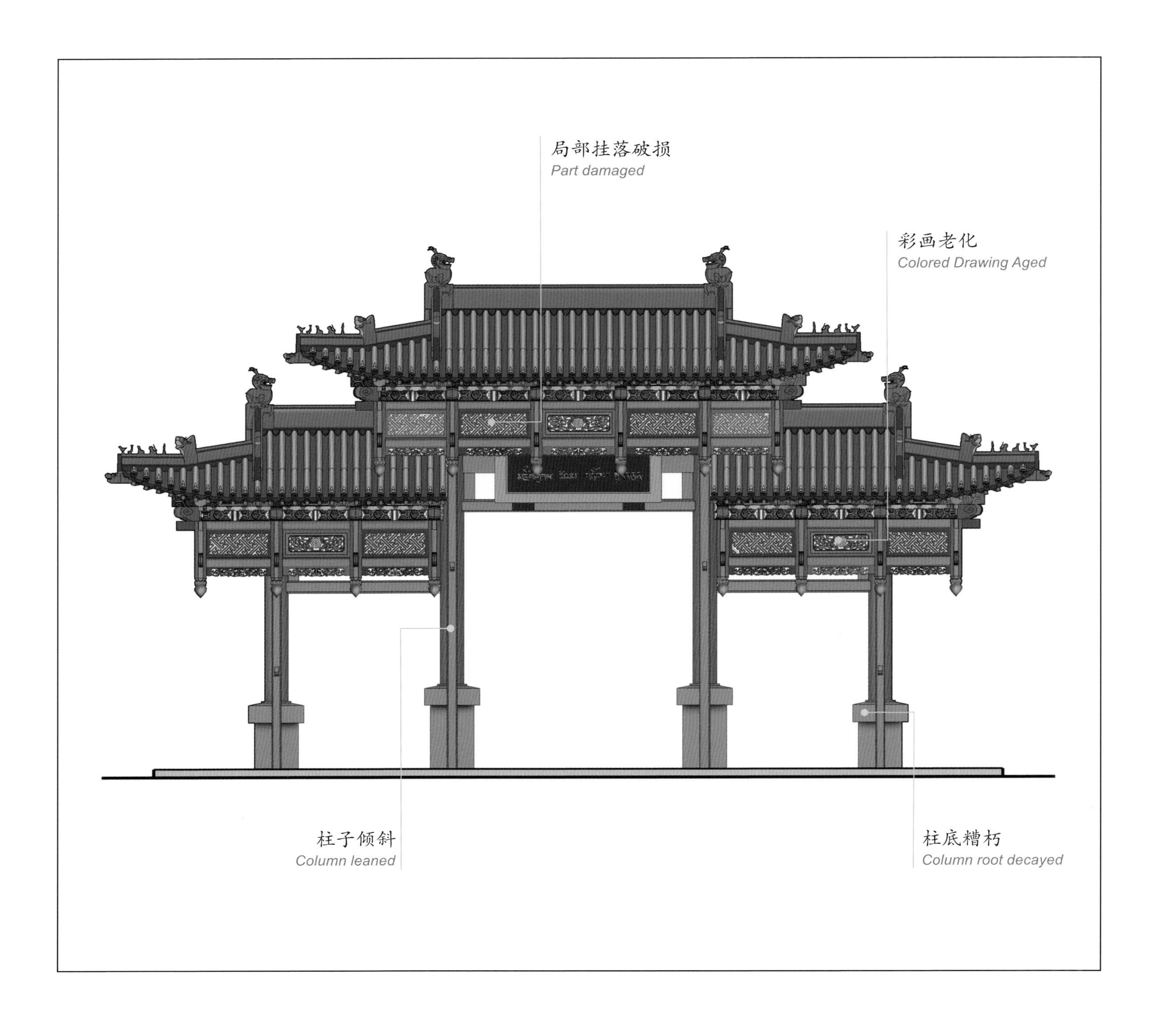

中牌楼正立面
Front Vertical Face of Middle Archway

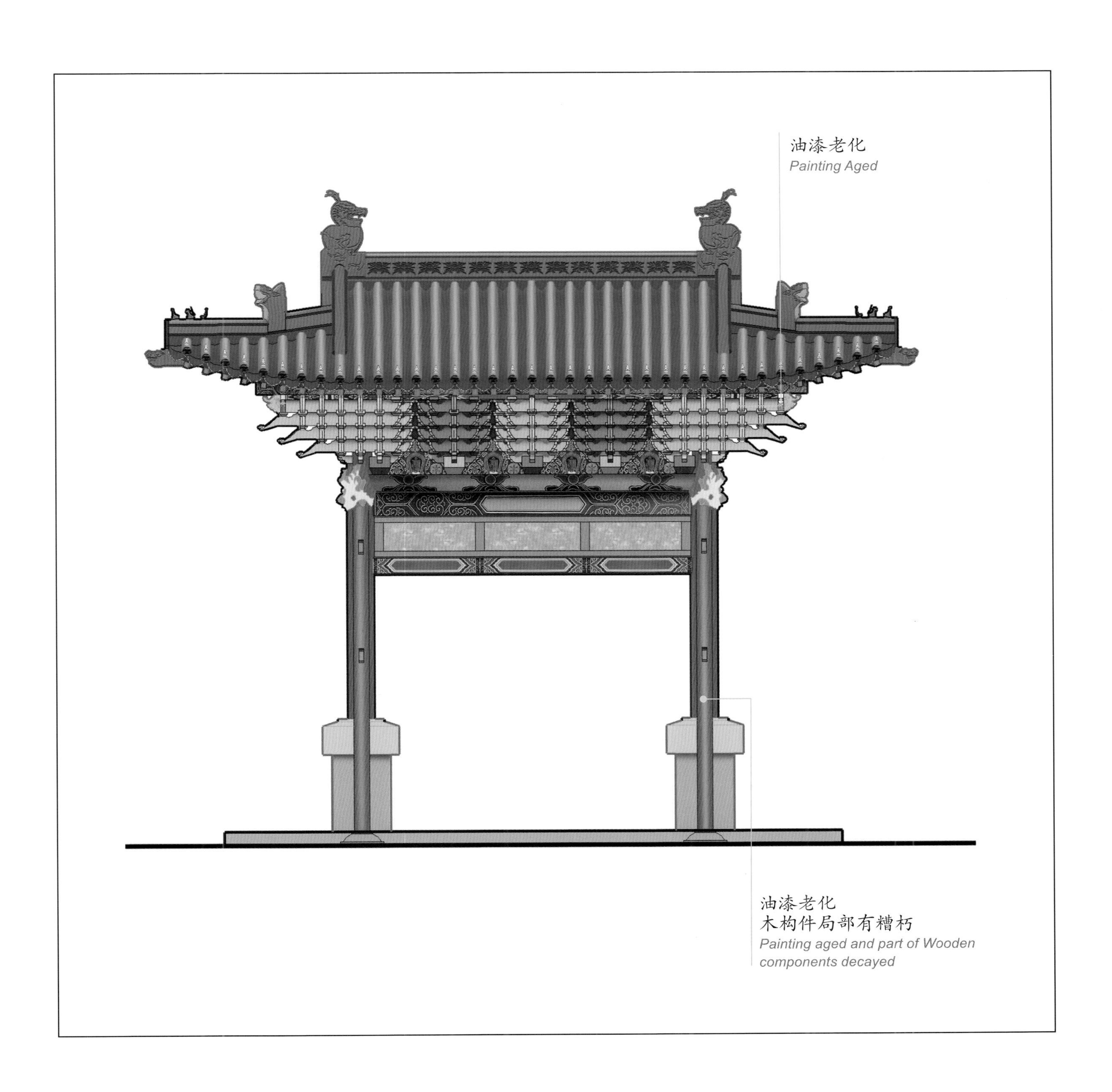

东西牌楼正立面
Front Vertical Face of east-west Archways

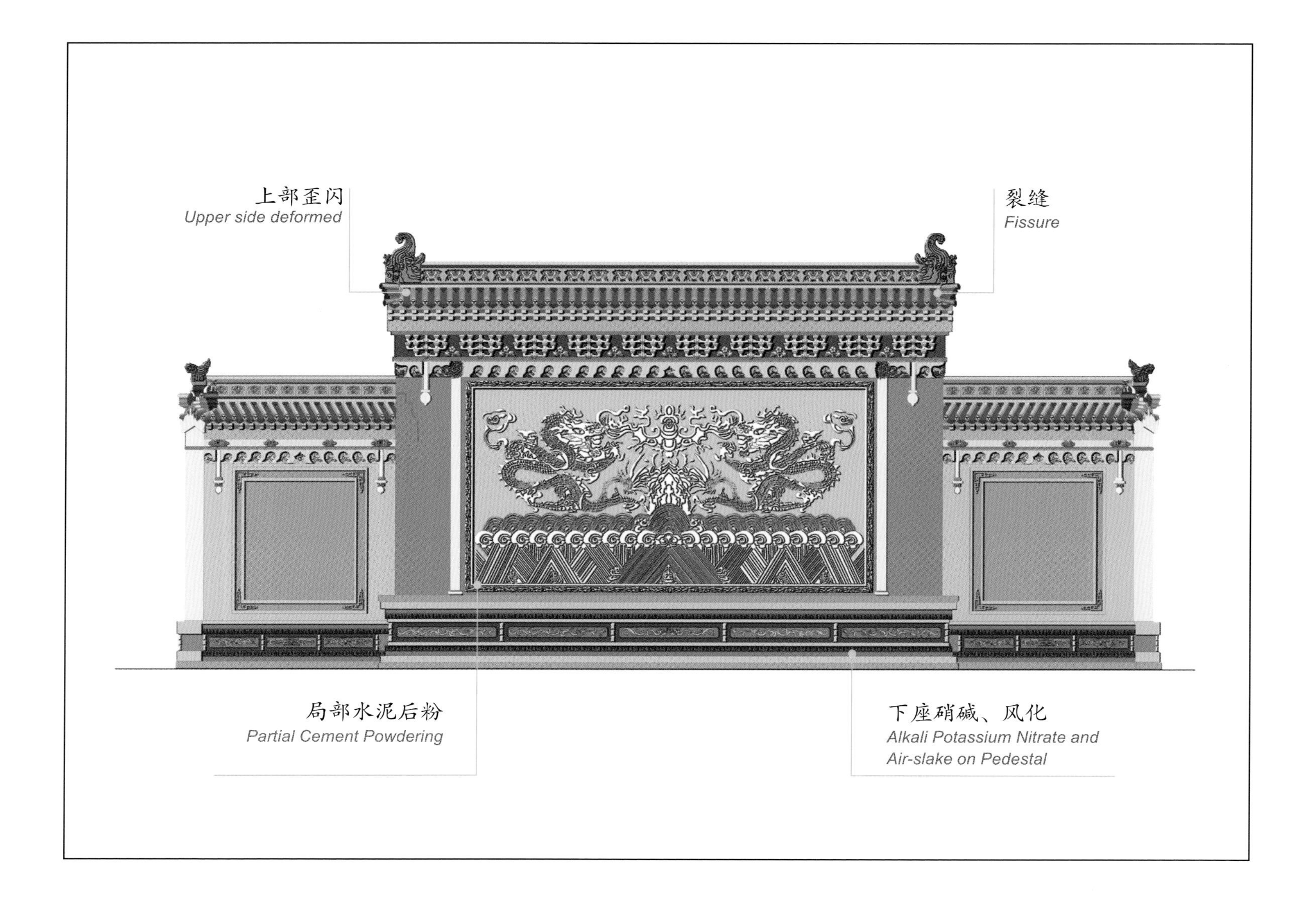

照壁正立面
Front Vertical Face of Screen Wall

大门地面下沉
Land subsidence

大门门扉扭曲
The Front Gate deformed

大门梁架变形下垂
Beam Frame deformed and sagged

大门屋面残损锈蚀
Roof Covering damaged and corroded

中牌楼柱根糟朽，混凝土夹杆石
Column root decayed, concrete Pole Clamping Plinth

中牌楼局部挂落缺失破损
Part absent and damaged

东牌楼油漆老化局部糟朽
Paint Coat aged and decayed

东便门博风板残破
Gable eave board damaged

照壁上部歪闪
Upper side deformed

照壁墙体裂缝
Wall Fissure

照壁后期维修水泥粉饰
Cement Powdering for Later period maintaining

照壁下座硝碱、风化
Alkali Potassium Nitrate and Air-slake on Pedestal

大门侧面倾斜
Flank side of the Front Gate leaned

西便门残破歪斜
Side door skew and damaged

中牌楼侧面倾斜
Flank side of Middle Archway leaned

宫墙倾倒，柱子糟朽
Palace wall dumped, Column decayed

2.2 维修设计

根据各单体建筑存在的不同问题，对各单体建筑维修设计方案分述于下：

大门：

大木构架：校正梁架，将出挑下垂部件恢复至水平位置，并用铁件紧固拉结；墩接下沉，开裂柱子；加固柱子基础；用石材夹杆石更换混凝土夹杆石。

屋面：更换严重锈蚀瓦件，其余瓦件除锈并重新刷漆。

墙体：重新批灰、油饰。

地面：重新铺设地面；增设台明、散水。

油饰彩画：部分大木构件重新油漆。

东、西便门及宫墙：

基础：开挖基槽，做砼垫层，将柱根与两侧的槽钢一起埋入其中，柱根部做了防腐处理，柱与柱之间做砌砖基槽。结合地面铺装抬升柱子；校正梁架；加固榫卯；添补缺失构件。

增设台明及地面铺装；添补柱础；增设散水。

化学法清洗保护彩画。

油饰屋面及大木下架、大门。

校正木围墙，添补、更换损坏、缺失围墙木构件；重新油饰。

牌楼：

揭除屋面；校正梁架，加固榫卯；墩接糟朽柱子、戗柱；更换糟朽仔角梁；修补损坏花板。

增设台明；重新铺装地面；更换混凝土夹杆石为石材夹杆石；增设散水。

化学法清洗保护彩画；油饰大木下架。

屋面除锈，重新油饰。

照壁：

加固基础：整体加固照壁基础。

拓换硝碱残破下碱面层砌体，下碱内衬砌体裂缝内灌注水泥浆，做整体加固。

水泥浆加建筑胶灌注弥合墙体及屋面裂缝。

瓦面刷隔水剂作防渗处理。

重新铺设散水。

门前区广场：

因各单体建筑均增设台明，应统一调整广场地面标高，依原道路格局重新铺设路面。

铺设停车场地面，增设广场绿化区域，重新设计广场排水走向，增设地下排水管网。

原大门、牌楼地面，门前广场道路均用条砖铺设，东、西便门无地面铺设。此次维修将大门、牌楼、东、西便门地面用方砖铺设，并增设台明。其余道路用条砖铺设。

2.2 Architecture maintaining plan design

According to different problems existing in each building, specific maintaining plans have been made as follows:

2.2.1The gate:

Wooden structure: correct beam frame, recover the sagged section to horizontal position and fasten it by iron piece; the connection part sagged, pillar cracked; strengthen the pillar base; replace the stony pole clamping plinth with concrete one of 1600×730×730.

Roof covering: renew the severe corroded tiles, dedust and repaint others.

Wall body: repave the lime and repaint the colored drawings.

The ground: repave the ground, platform and drainage platform.

Wooden structure: correct the deformed gate.

Colored painting: repaint the under-beam and the back of the front gate.

2.2.2East-west side doors and palace wall:

Ground work: excavate foundation ditch to 800 wide, pave concrete bedding of C20, build 600＊600 concrete pier for each column, embed column root along with channel beam into concrete pier to 600 deep, do anticorrosion treatment for column root, pave bricks at foundation ditch between each column.

Upraise column according to the ground elevation; rectify the beam frame, strengthen the mortise and tenon joint, supplement absent components.

Repave the platform and the ground; rebuild the column plinth; construct drainage platform.

Chemical cleaning and protection are needed for colored paintings.

Repaint the roof covering, under-beam wooden structure and the front gate.

Correct the wooden bounding wall, rebuild the damaged and absent wooden structure; repaint the wall.

2.2.3Decorated archway

Remove the roof covering; correct beam frame, strengthen the mortise and tenon joint, connect the decayed column and heavy pillar; replace decayed hip rafter; repair the damaged painted boards.

Pave the platform; repave the ground; replace the concrete pole clamping plinth with stony one, build drainage platform.

Chemical cleaning and protection for colored paintings; repaint under-beam wooden structure.

Remove the rust on roof covering and repaint it.

2.2.4Entrance screen wall

Strengthen the screen wall foundation integrally.

Replace alkali layer on the wall, fill the fissures with cement paste, strengthen the whole wall body.

Fill the damaged wall and fissures in roof covering with mixture of cement paste and building sealant.

Brush the waterproof material on tiles.

Repave the drainage platform.

2.2.5The front square:

Adjust the ground elevation due to newly built platform, repave the ground according to former ground situation.

Pave the parking ground; expand afforested area.

Redesign drainage direction at the square, build underground drainage pipe network.

Former front gate, ground at archway and the road at front square are paved by bricks, no ground laying at east-west side doors. For differentiating the construct ground and road ground, and rank difference between construct ground and road ground, the ground at front gate, archway, east-west side doors are paved by bricks.

大门测绘
Gate surveying and mapping

专家视察
Experts Inspecting

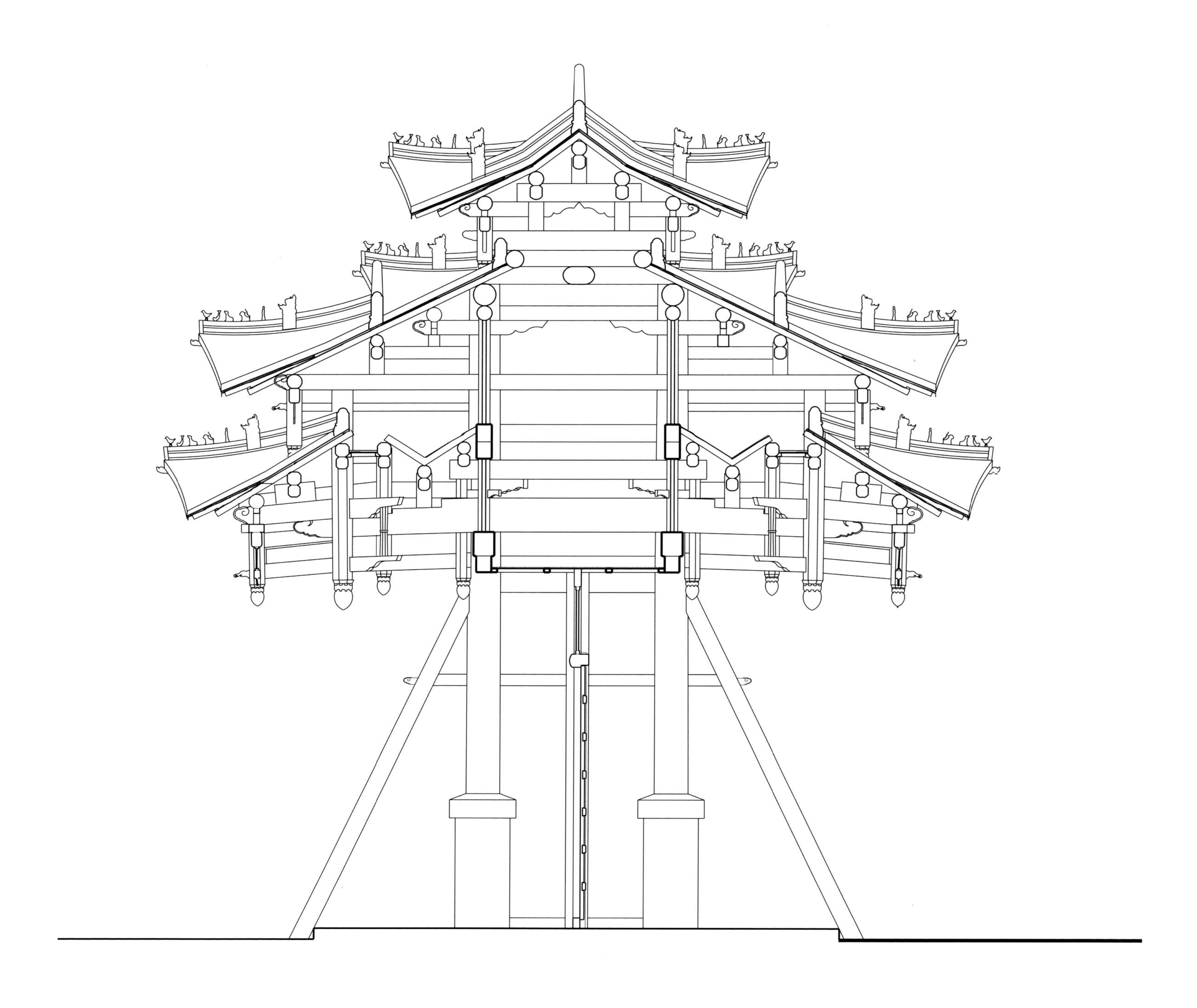

大门剖面图
Profile Map of the Front Gate

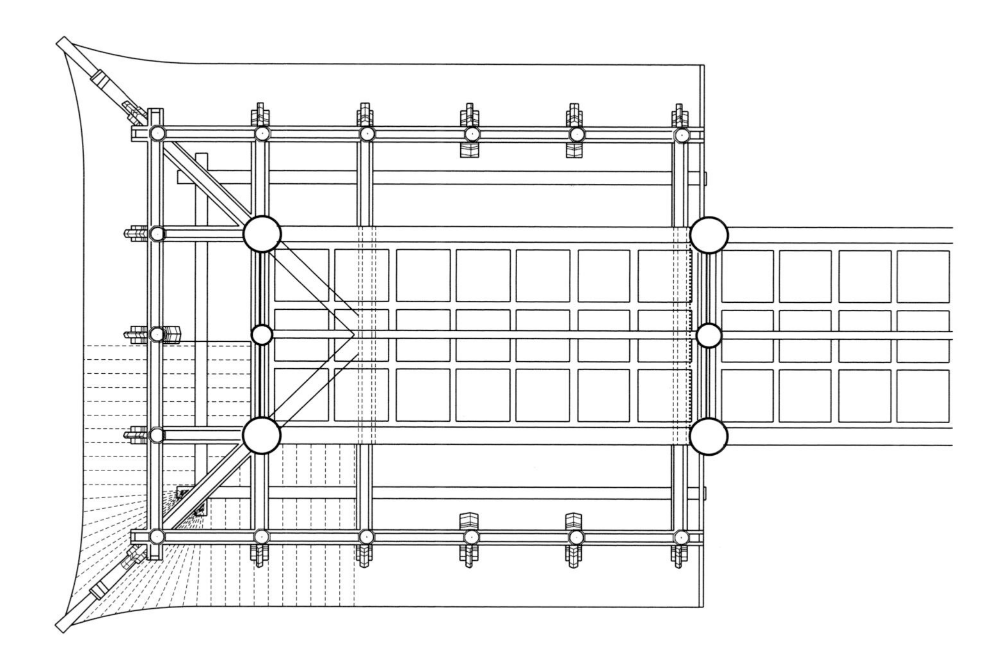

大门次楼一层梁架仰视图
Upward View of Beam Frame at first floor of Subordinate Building Gate

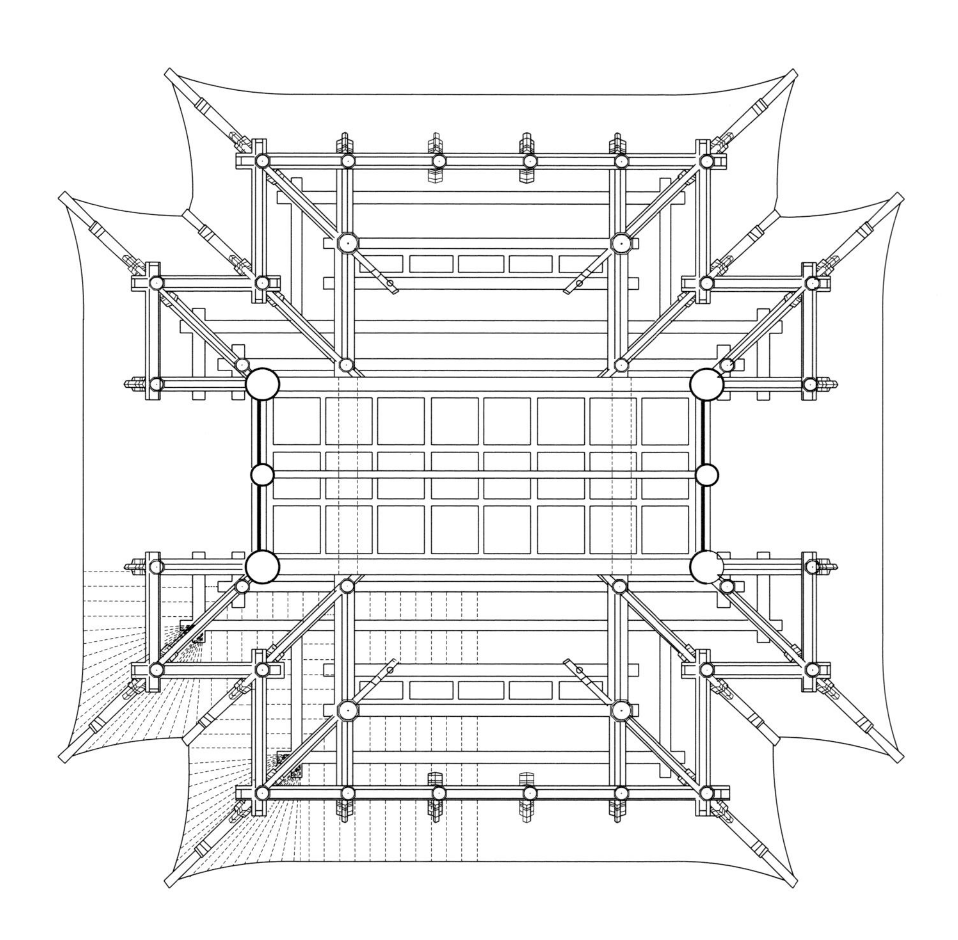

大门主楼一层梁架仰视图
Upward View of Beam Frame at first floor of Main Building Gate

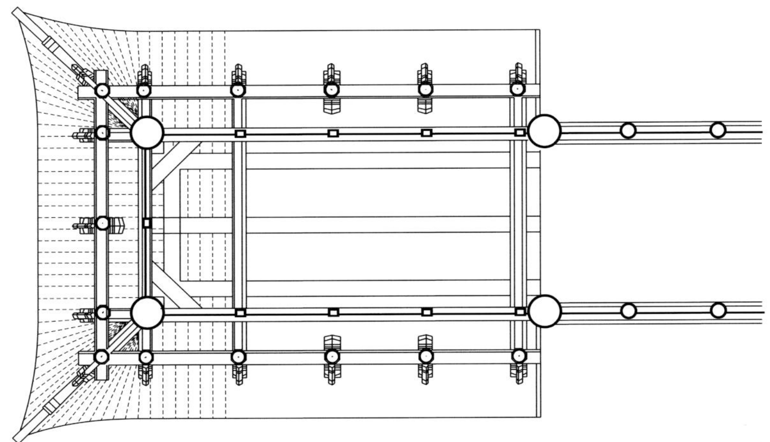

大门次楼二层梁架仰视图
Upward View of Beam Frame at second floor of Subordinate Building Gate

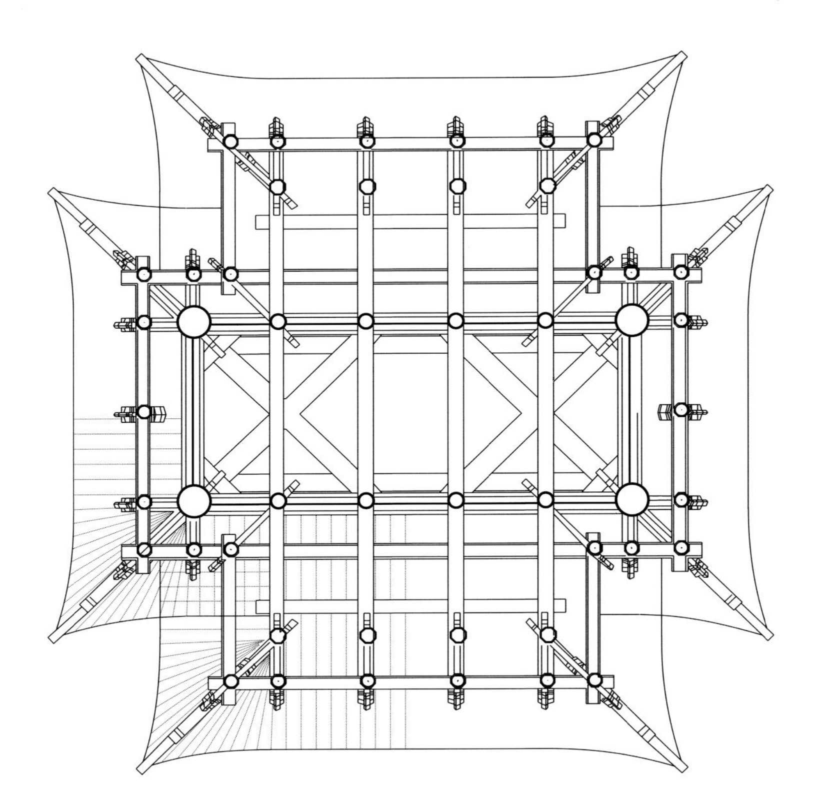

大门主楼二层梁架仰视图
Upward View of Beam Frame at second floor of Main Building Gate

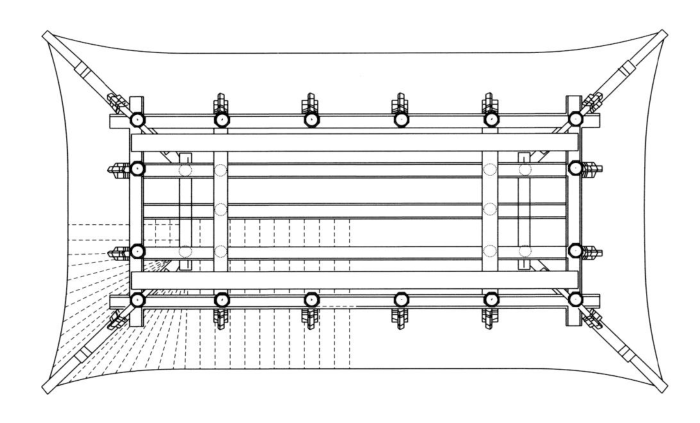

大门主楼三层梁架仰视图
Upward View of Beam Frame at third floor of Main Building Gate

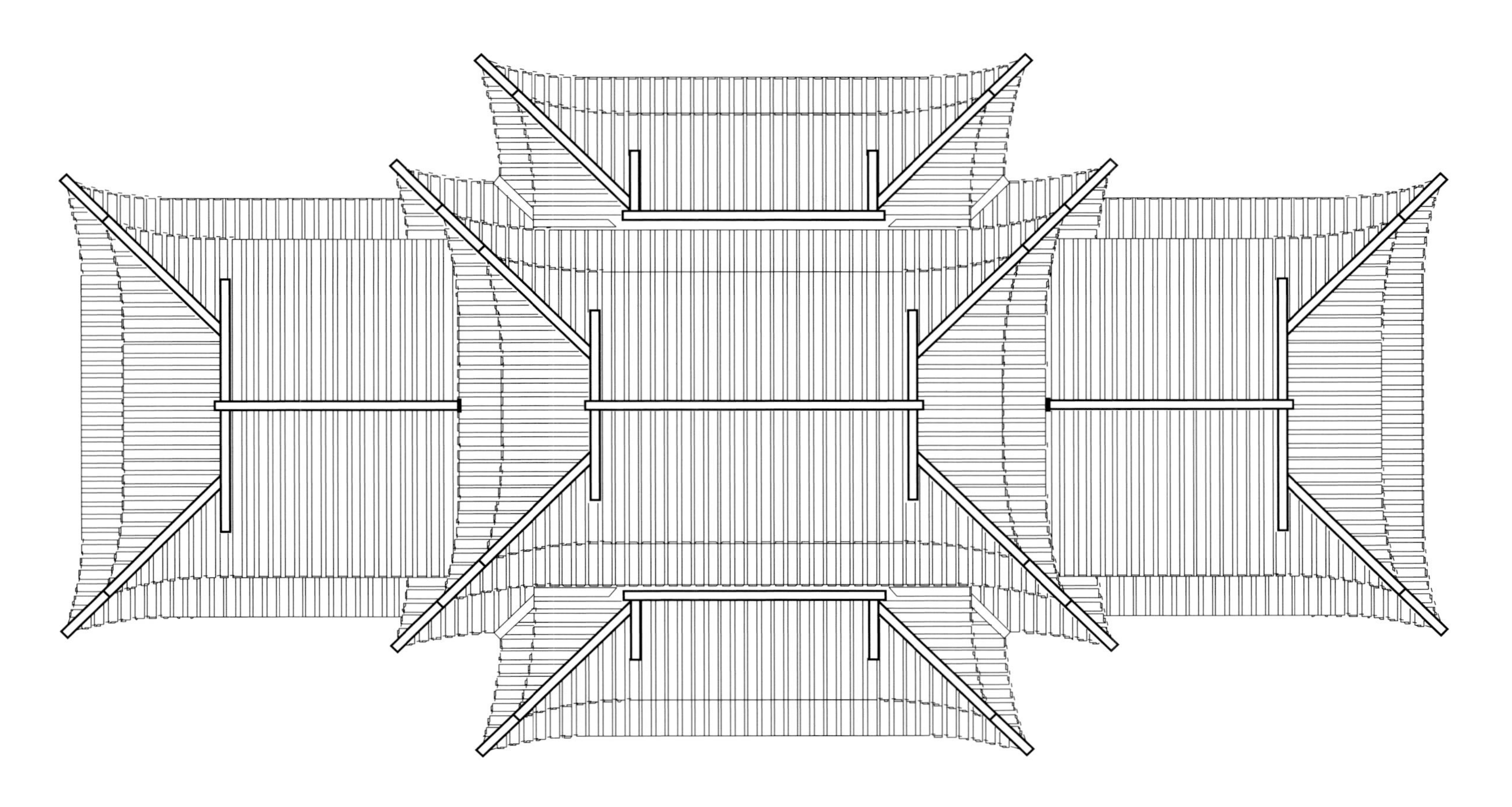

大门屋面俯视图
Top View of Roof Covering on Front Gate

大门正立面图
Front Elevation of the Gate

大门侧立面图
Side Elevation of the Gate

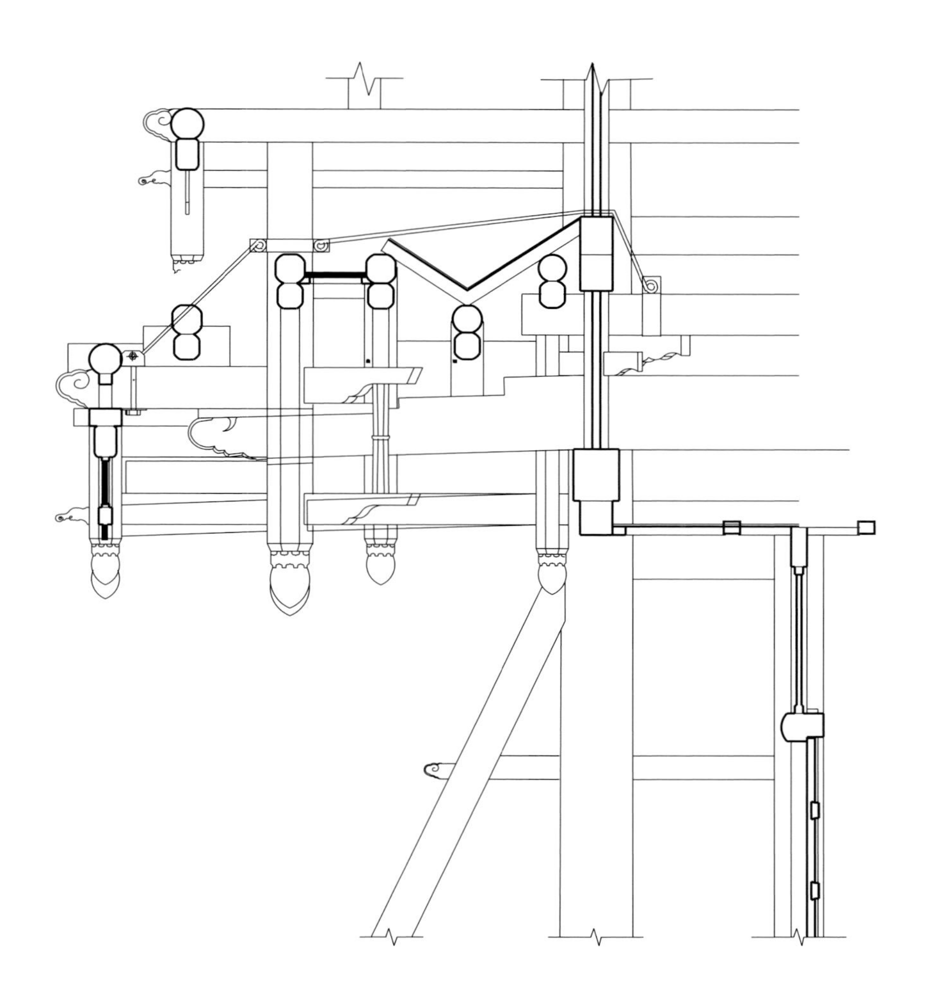

大门主楼一层梁架校正示意图
Correction Sketch Map of Beam Frame at first floor of Main Building Gate

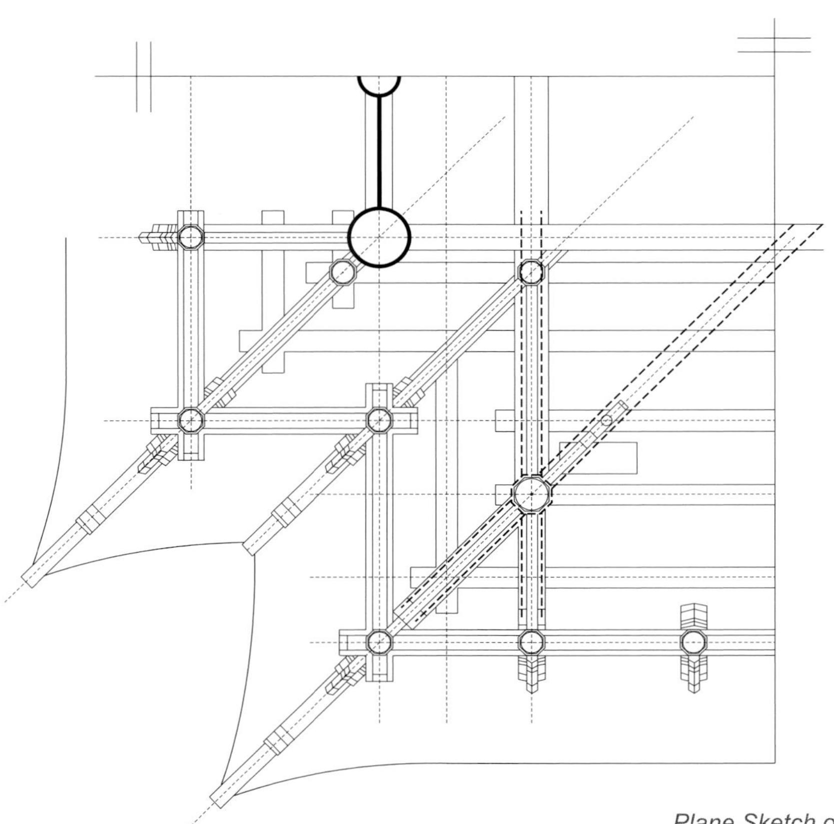

大门钢索布置平面示意图
Plane Sketch of Gate Wire Rope Arrangement

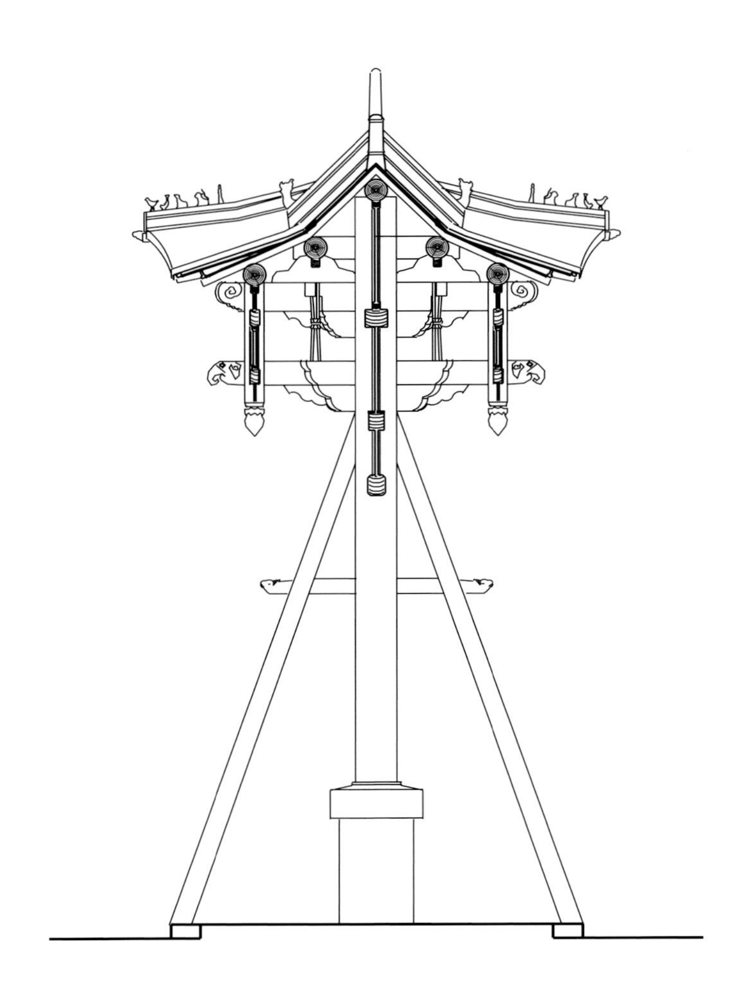

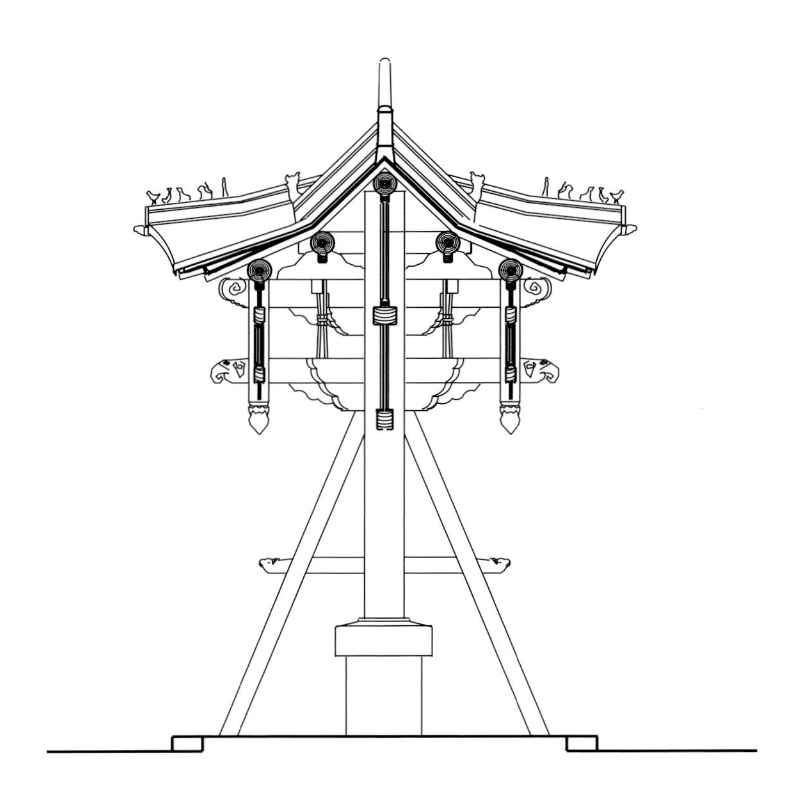

中牌楼剖面图
Profile Map of Middle Archway

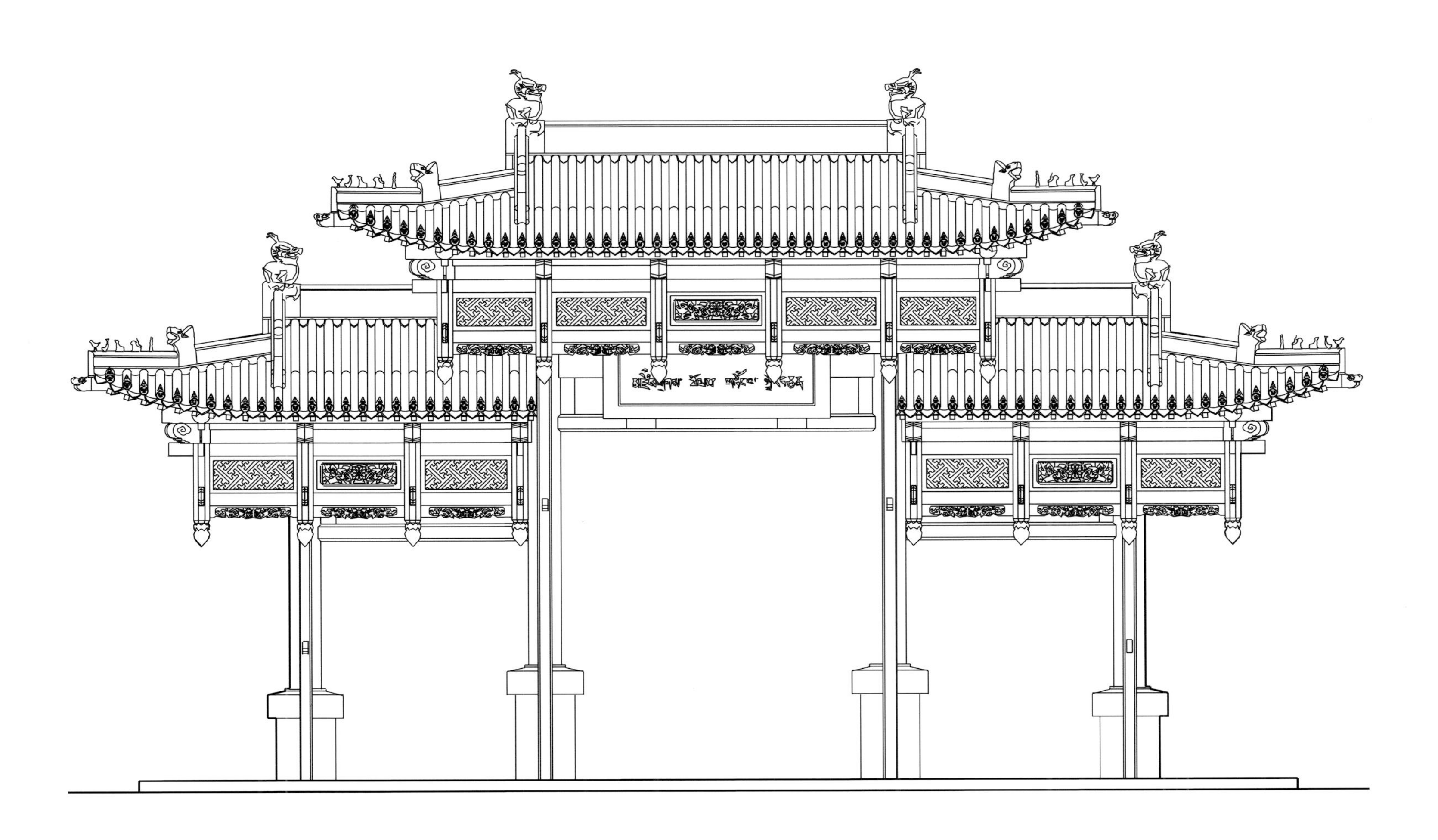

中牌楼南立面图
South Vertical View of Middle Archway

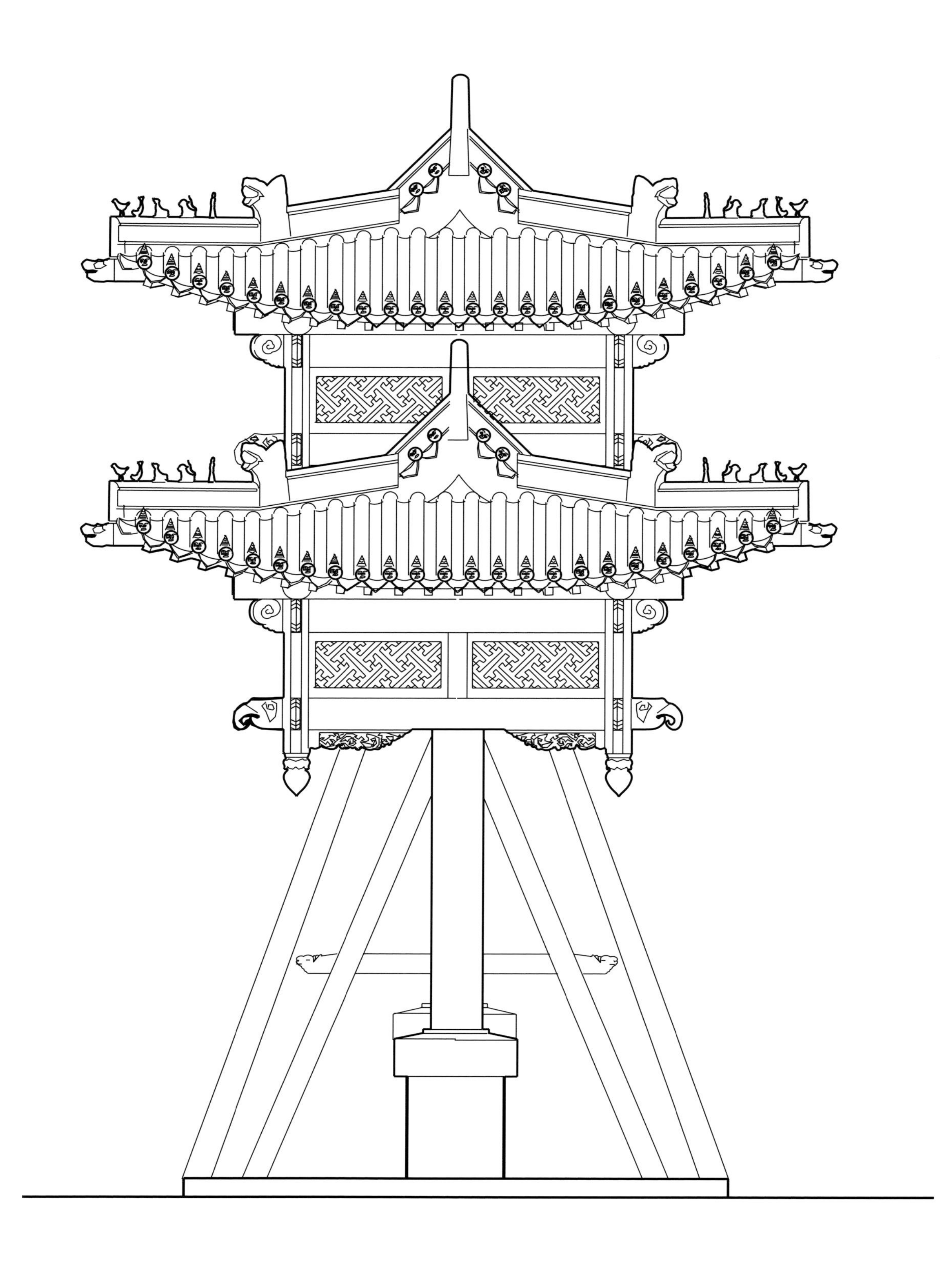

中牌楼侧立面图
Side Vertical View of Middle Archway

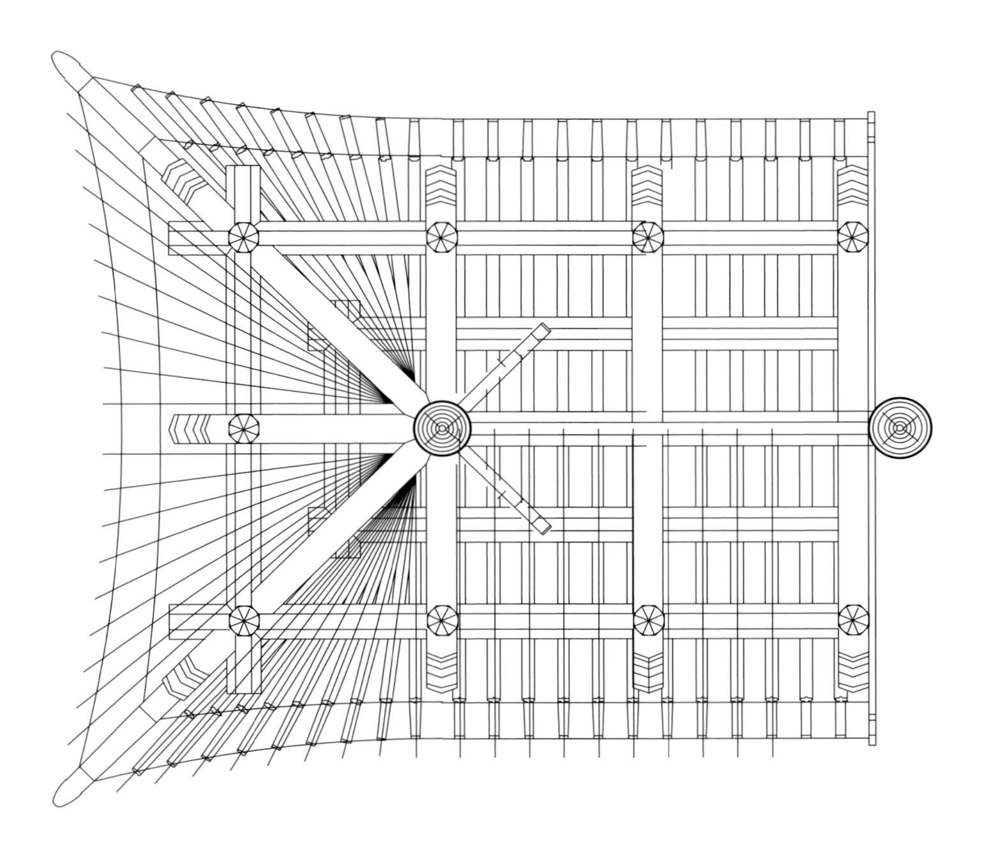

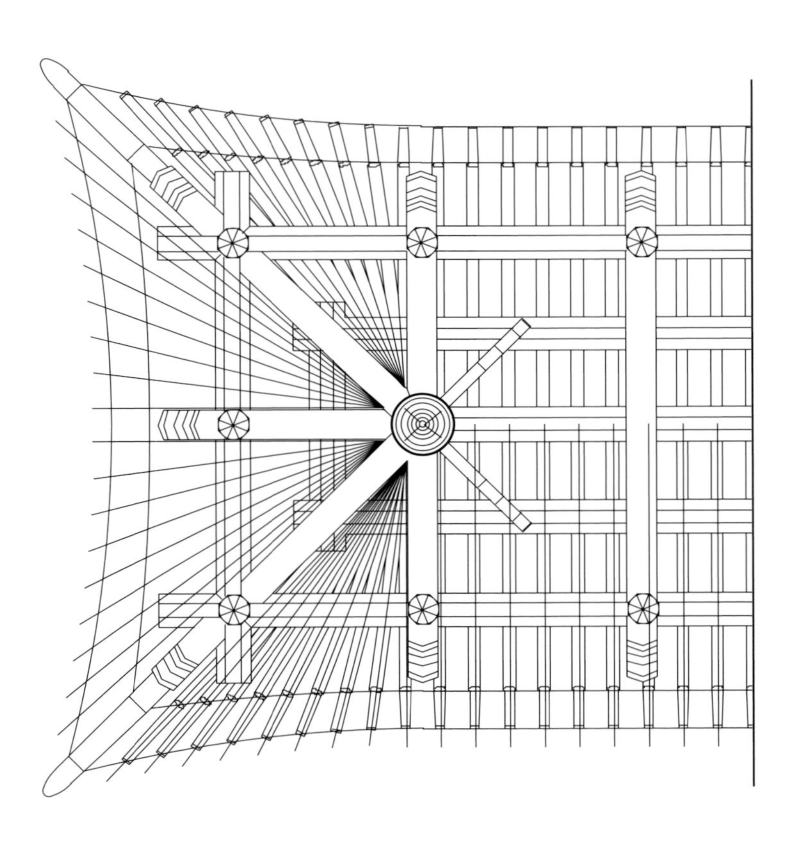

中牌楼梁架仰视图
Upward View of Beam Frame on Middle Archway

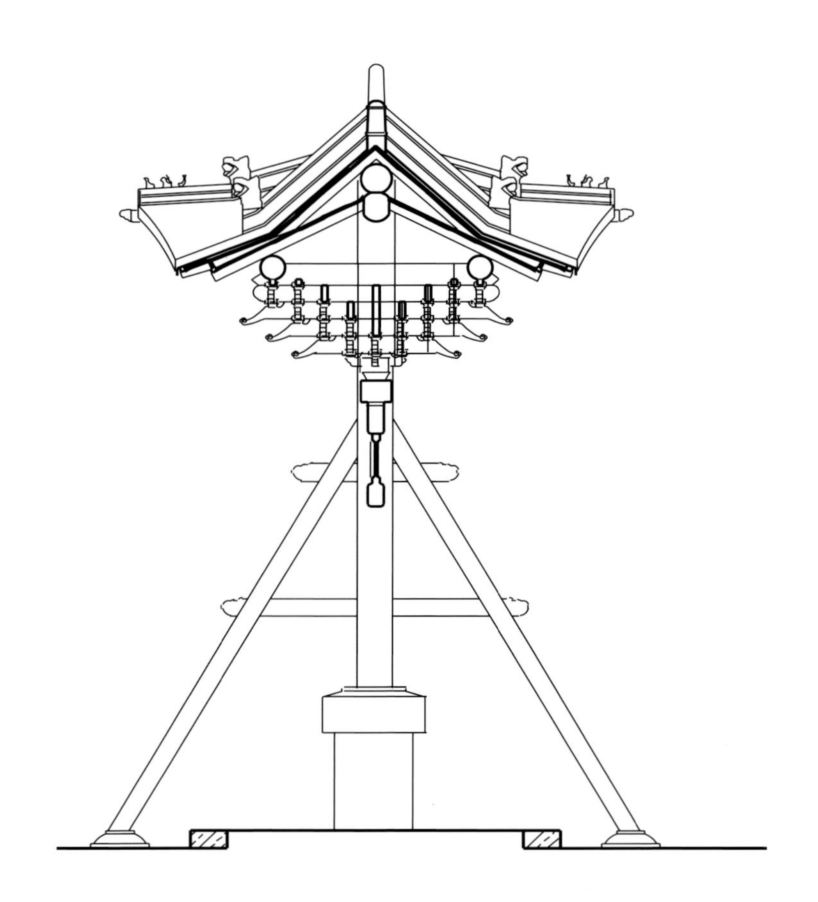

东西牌楼剖面图
Profile Map of east-west Archways

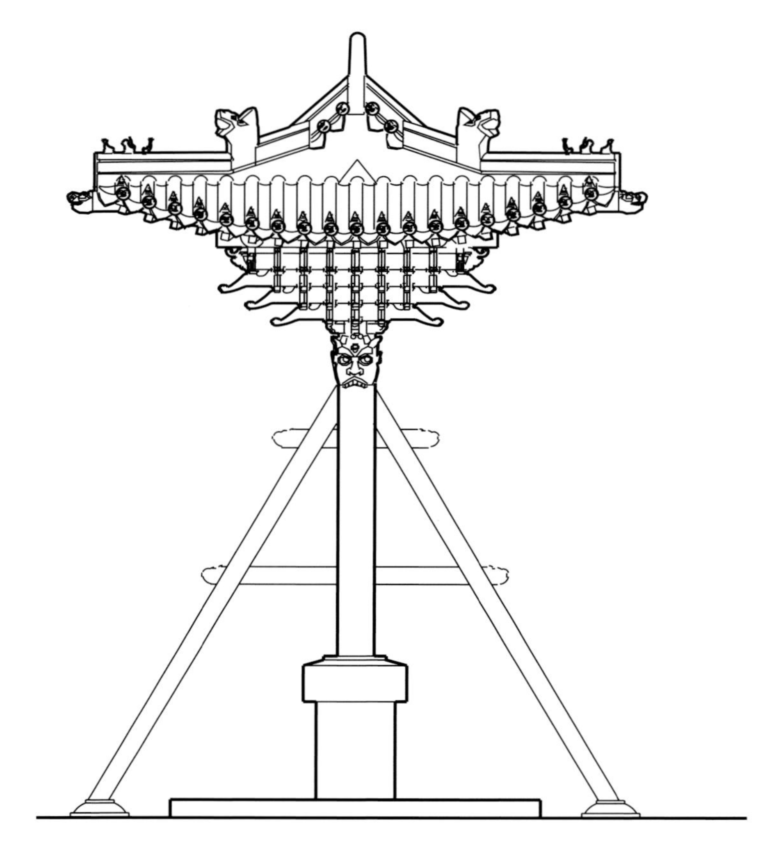

东西牌楼侧立面图
Side Vertical View of east-west Archways

东西牌楼立面图
Vertical View of east-west Archways

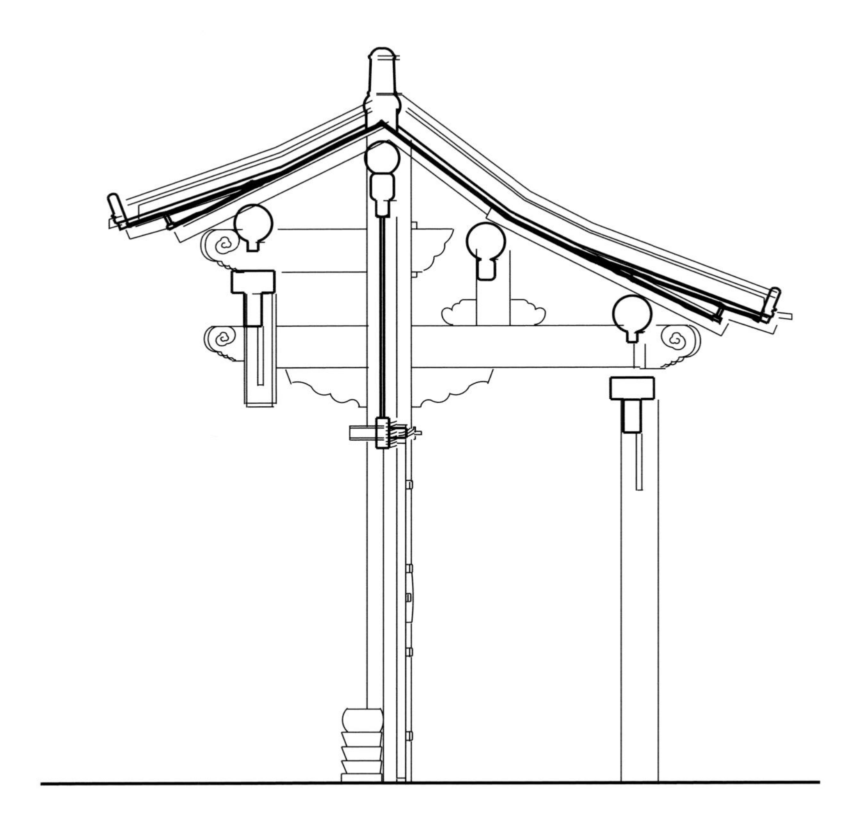

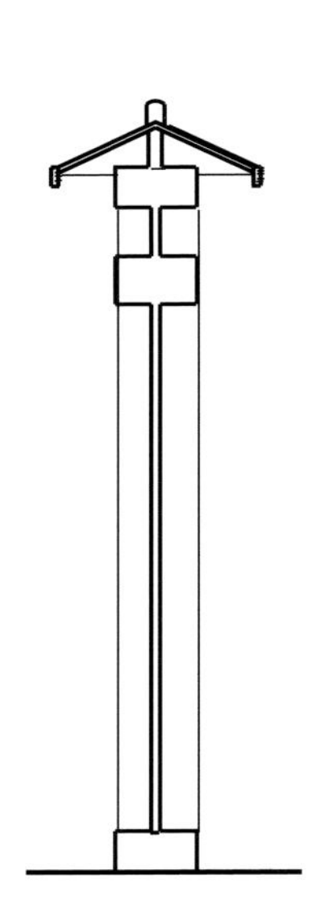

东西便门剖面图
Profile Map of east-west Side Doors

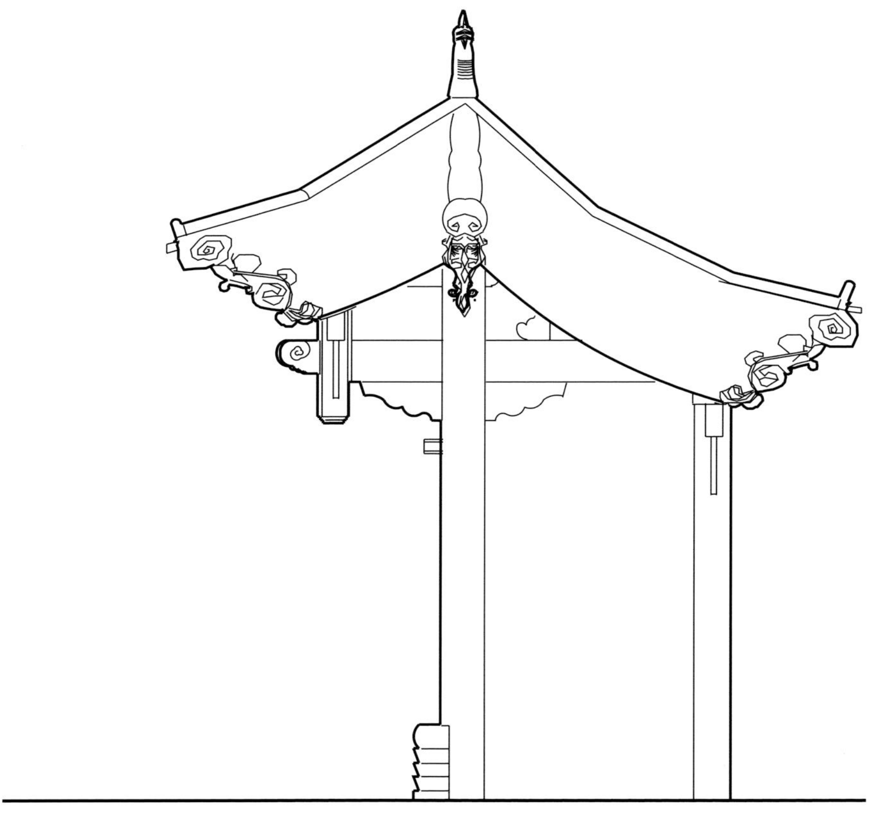

东西便门侧立面图
Side Vertical View of east-west Side Doors

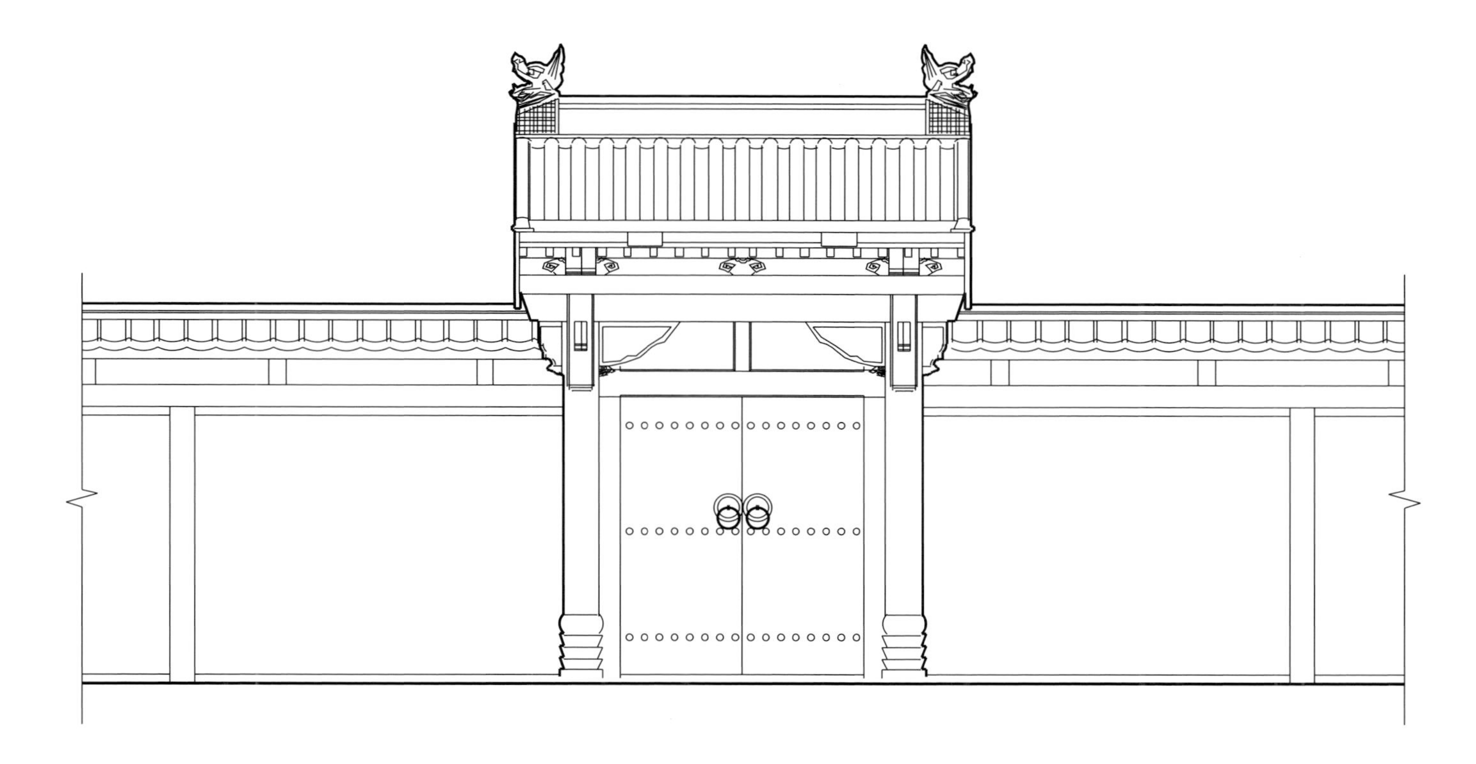

东西便门南立面图
South Vertical View of east-west Side Doors

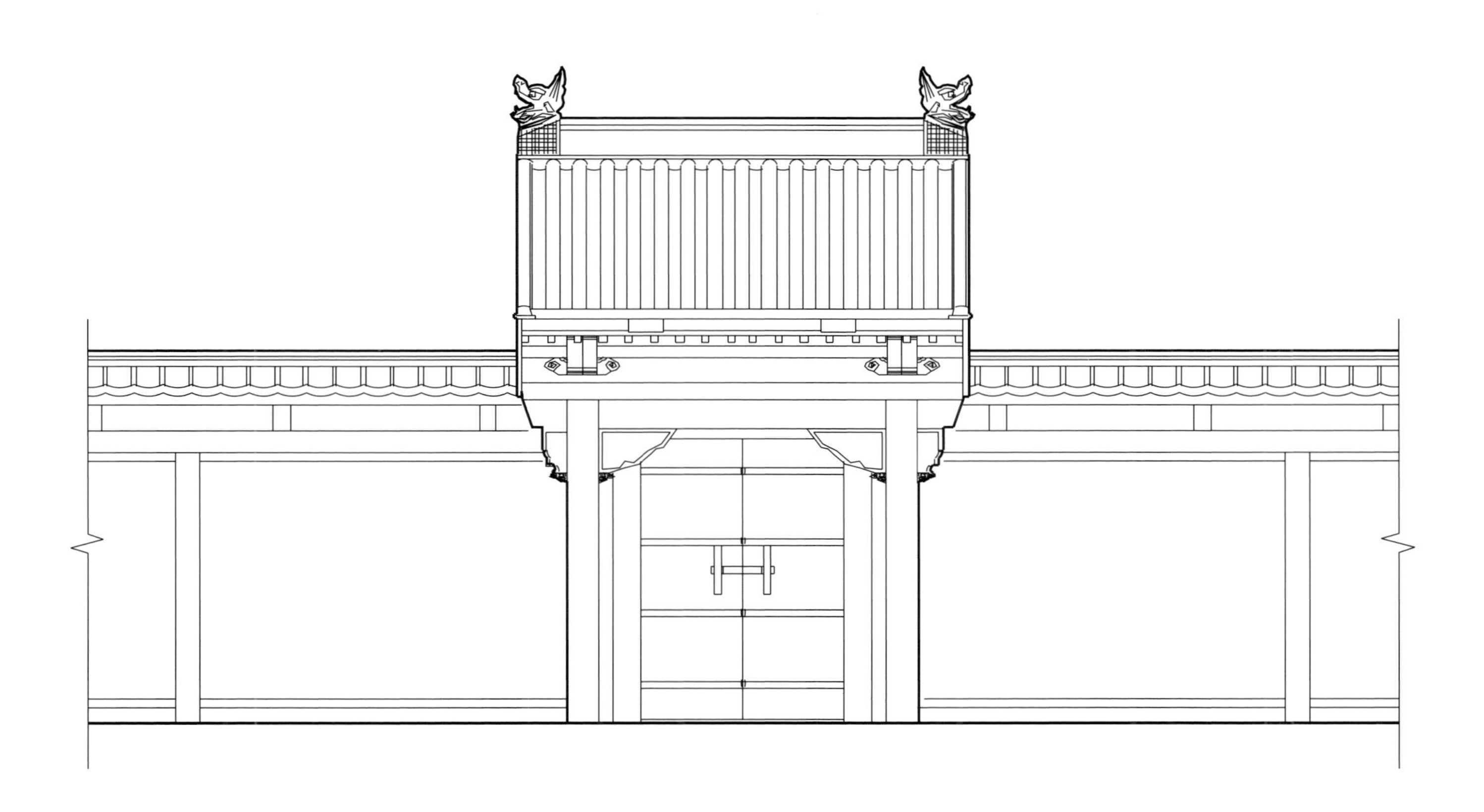

东西便门北立面图
North Vertical View of east-west Side Doors

照壁侧立面图
Side Vertical View of Screen wall

照壁剖面图一
Profile Map of Screen wall (1)

照壁剖面图二
Profile Map of Screen wall (2)

照壁北立面图
North Vertical View of Screen wall

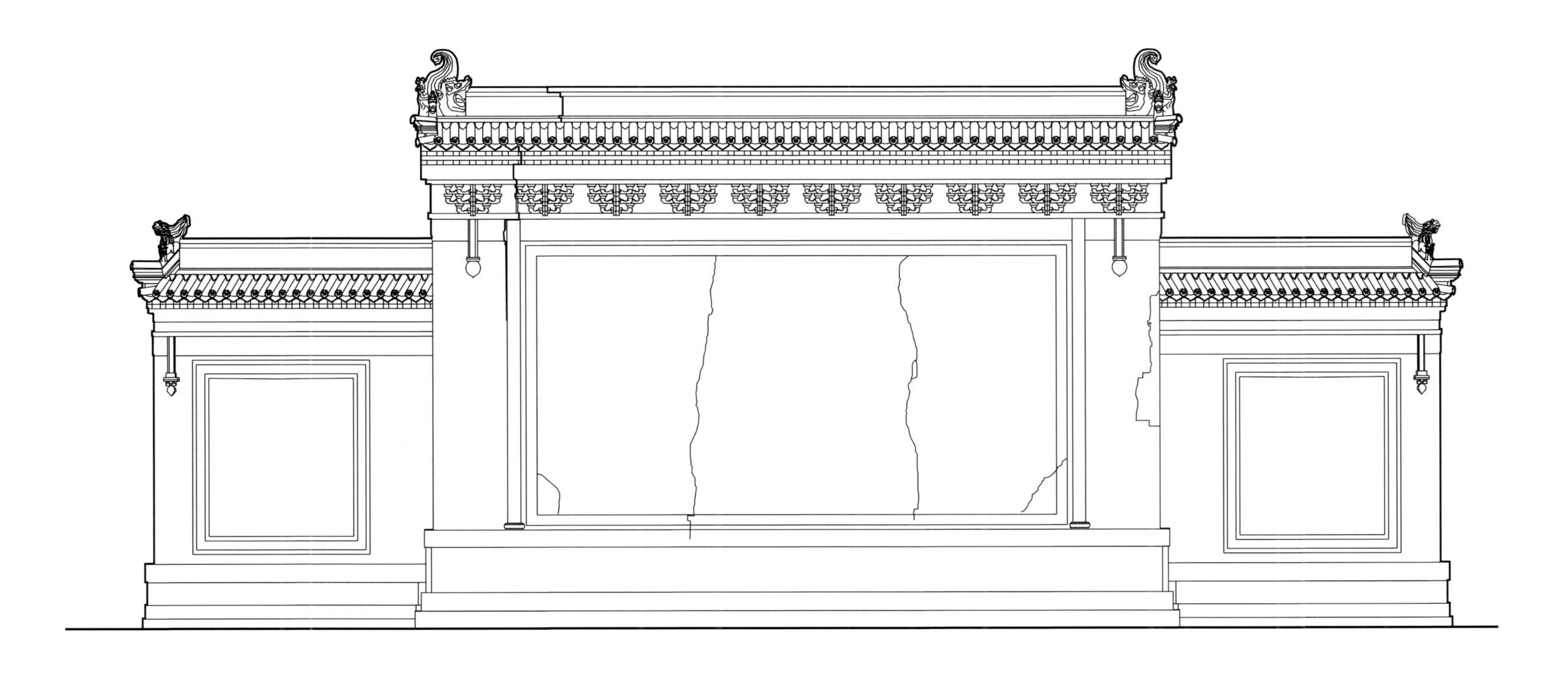

照壁南立面图
South Vertical View of Screen wall

2.3 本体的维修实施

博格达汗宫博物馆自建立以来一直没有进行过大的维修，总体形制统一，各单体外观保存基本完好，但均存在一定的隐患，因各种因素作用出现不同程度的变形、残损。博格达汗宫门前区建筑的维修自2006年5月21日进场，9月22日离场，2007年4月25日进场，9月30日竣工，共历时17个月。2006年完成大门、东西便门、南宫墙建筑的维修；2007年完成中牌楼、东西牌楼、东西旗杆、照壁、木栅栏及院内环境的整治。在此期间，国家文物局组织专家组进行了检查，各位专家一致认为维修符合保护原则，现场组织管理规范，已经达到设计的目的。与此同时，蒙古国政府官员、相关专家通过到现场的监督检查，高度评价了维修的效果。2006年9月20日，由大使馆组织召开了新闻发布会，共有蒙古9家媒体到现场进行了采访和专访，博物馆馆长通过各方面的对比，对双方的合作十分满意，对维修组织和维修手段高度赞扬，认为维修的效果非常好。

在维修过程中，我们主要对各单体进行了以下维修内容：

2.3.1 大门维修内容

校正梁架；墩接下沉，开裂柱子，加固柱子基础，更换混凝土夹杆石；屋面更换严重锈蚀瓦件，其余瓦件除锈并重新油漆；大木构件修补地仗、重新油饰；地面：重新铺设地面；增设台明、散水；木装修校正走闪大门。

2.3Architecture maintenance

Bogd Khaan Palace Museum has never undergone vast maintenance since establishment, the overall shape and structure is unitive, each section is preserved well, however, distortion and damage gradually emerge because of various kinds of reasons, hidden problems still exist. The front square maintenance project of Bogd Khaan Palace Museum started on May 21st 2006 and ended on Sep.22nd in the same year, and then started on Apr. 25th 2007 again and finished on Sep. 30th which lasted for seventeen months. The maintenance project on front gate, east-west side doors and south palace wall were completed in 2006; the middle archway, east-west archways, east-west flagpoles, entrance screen wall, wooden stockade and the surroundings inside the yard were fixed in 2007. For the mean time, the state administration of cultural heritage organized experts group to inspect the project site, finally, the maintenance effect satisfied these experts who considered it was conformed with protection principles and organization management regulations which attained the design aim. Officials from Mongolia government and experts highly praised the maintenance effect through field inspection. Press conference was held by embassy on Sep. 20th 2006, nine Mongolian media made interviews or exclusive interviews, curator of the museum expressed his satisfaction on Sino-Mongolia’ s cooperation and spoke highly of the maintaining organization and means, considering that the effect was beyond compare.

Sections mainly being maintained during project implementation are as follows:

2.3.1Details of gate maintenance

The following works should be done: Correction of beam frame, the sinking decayed foot replaced by wooden column, cracked pillar, consolidation of column base; replacement of concrete pole clamping plinth and severely corroded tiles on the roof, other tiles should be derusted and repainted; the base layer on wooden components must be repaired and repainted; the ground should be repaved; the platform and drainage platform around buildings should be added and the deformed gate should be corrected.

搭架与支撑

Erection of frame and jackstay

基础的处理

大门和三座牌楼原来为混凝土夹杆石，此次维修更换为石质夹杆石。开挖后发现，原来的柱坑是由杂土回填，且无夯实，因此产生了下陷现象，使人误以为基础出了问题。 此次维修保持了原有做法，在柱基处只是增加了砼基础垫层。检查戗柱根部，其基础为混凝土浇注。本次维修增加了花岗岩柱础石，使戗柱根部插入在石基中，不再埋入地下。根据柱子糟朽程度不同进行维修，对糟朽严重的柱子进行墩接，对表皮糟朽约3厘米以内的柱子采用剔补。在墩接主楼西北柱时，发现留有蒙文，表明1989年曾进行过维修。

为了隔断土壤与柱子直接接触，在这次更换夹杆石的过程中，将埋入地下段沿袭以前做法，即将砼夹杆石截短埋入地下，地表之上使用花岗石夹杆石。夹杆石分为上下两段，上段为顶盖，由两块组成，下段为夹杆柱体，两块组成。在柱坑回填中，将原杂土过筛，而且掺入灰土，用人工夯实。

Foundation maintenance

The pole clamping plinth at the front gate and three decorated archways previously were made of concrete, later they were replaced with stony ones. The former pillar hole was backfilled with rough soil without being tamped which resulted in sinking and generating misunderstanding about the foundation. The current maintenance keeps the previous means which add mercury bed course to the column base. When checking the root of props which serve as side support, it is found that the foundation is poured with concrete, granite plinth is introduced in this project which make the root of pillars inserted in the stony plinth instead of embedding underground. Considering different rotten degrees of pillars, connection works have been done for those severely rotten pillars, for pillars decayed from surface into deep side within 3 cm, the mean of scraping and patching is adopted. Mongolian language carved on north-west pillar of the main building was found when maintaining, which indicated that the repairment also happened in 1989.

To avoid the direct contact between soil and pillar root, the portion underground follows former ways of embedding concrete pole clamping plinth which has been cut short into the ground and for the portion above ground level, granite one is used. Stony pole clamping plinth can be divided into two sections, the upside is head cover which consists of two parts, the underpart is the pillar body consists of two parts too. In the process of pillar hole backfilling, the rough soil should be sieved first and then mixed with spodosol, finally tamped by manpower.

维修后
Condition after Maintaining

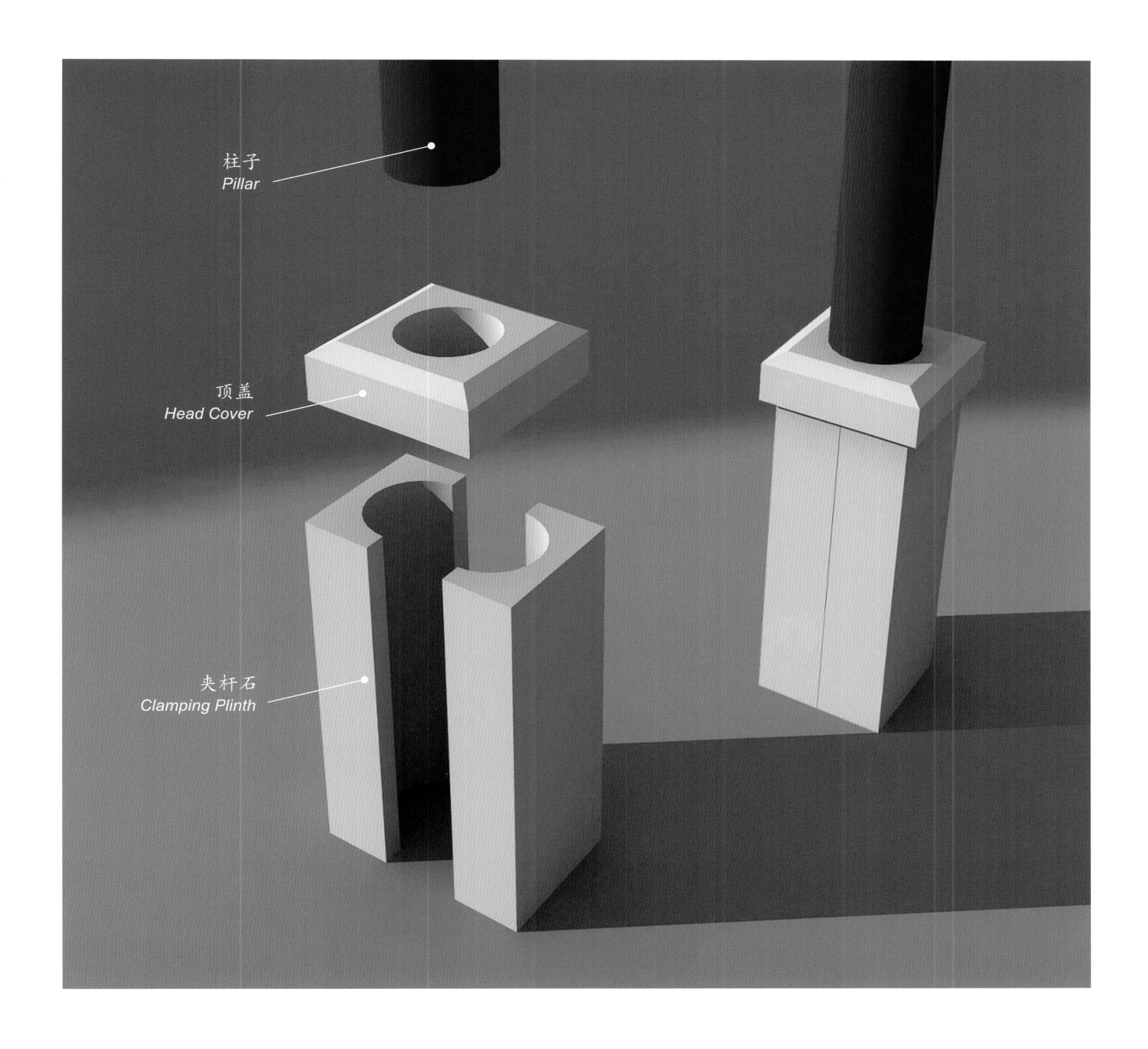

夹杆石构造做法
Structure of Pole Clamping Plinth

1989年大门柱子墩接题记
Inscription of Gate Pillar Connection in 1989

1961年7月大门维修题记
Inscription of Gate Maintenance in Jul. 1961

1989年墩接柱子糟朽状况
Condition of Decayed Pillar in 1989

2006年大门柱子墩接题记
Inscription of Gate Pillar Connection in 2006

铁带加固
Pillar Consolidation by iron strip

剔补柱子
Pillar Scraping and Patching

戗柱加箍，增设柱础
Prop hooping and Column Base adding

屋架、梁架的纠偏

通过测量，大门和中牌楼屋架、梁架都有不同程度的倾斜和下垂。大门主楼屋架向北倾斜18－20厘米，南梁架下垂14厘米，北梁架下垂7厘米，东西次楼一层檐口均有不同程度的下垂现象，中牌楼屋架向南倾斜14－16厘米。对于现状采用“牵、顶”之力进行纠正。首先在距主楼南柱12米处埋设地锚，使用倒链2个，地面斜撑杆6根，千斤顶2个，钢绳上部通过北柱缠绕在南柱上，再挂入倒链。纠偏时，倒链、千斤顶、撑杆同时进行，北撑杆加楔，南撑杆放松，待线锤到指定的位置后便停止动作，然后夹紧所有撑杆，使其稳定一段时间后再放松倒链、千斤顶。

Correction of roof truss and beam frame

The measurement data indicates that the front gate and beam frame of middle archway show different degrees of leaning and sagging. The roof truss of main building leans 18-20cm to the north, the south beam frame sags 14cm and the north one 7cm, sagging also can be seen at the first floor cornice of east and west subordinate buildings, additionally, the roof truss of middle archway leans 14-16cm to the south. These problems above should be corrected by dragging and pushing power. Firstly, earth anchor should be embedded 12m away from south pillar of main building with two inverted tooth chains, six jackstays and two lifting jacks, the upside of the steel sling should be twined from north pillar to south pillar and then inserted into inverted tooth chain. The adoption of inverted tooth chain, lifting jack and jackstay should start simultaneously, the wedge is added to north jackstay while loosening the south pole, the work should stop when plumbline shifts to specific location, then clamp all supporting poles for a while, finally loosen the inverted tooth chain and lifting jack.

钢索校正纠偏

Correction work by Steel Sling

斜撑校正纠偏

Correction work by Jackstay

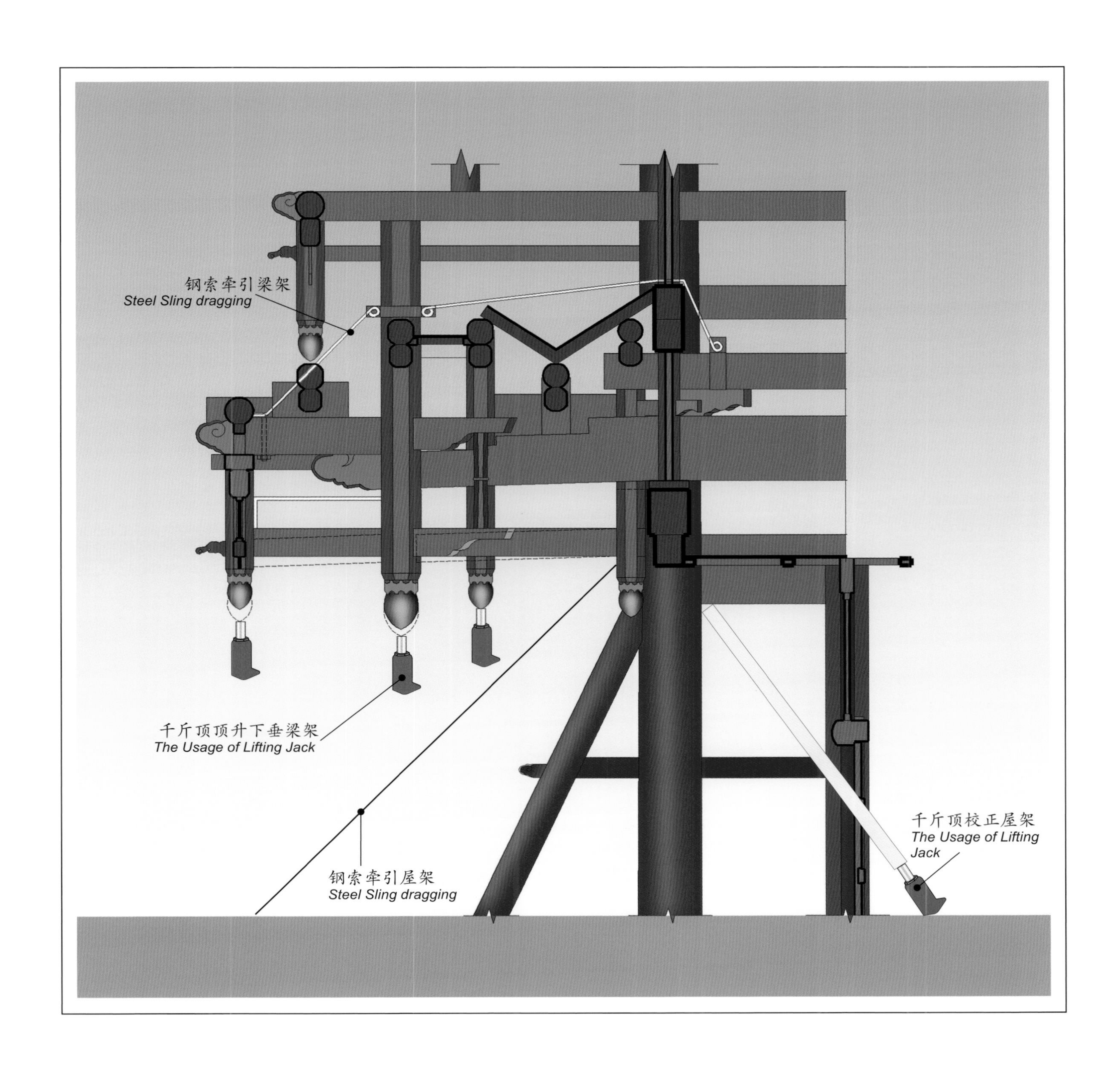

大门屋架、梁架校正示意模型
Model of Gate Roof Truss and Beam Frame Correction

大门主楼梁架变形
Beam Frame Deformation at Main Building

钢索牵引下垂梁架
Steel Sling dragging

千斤顶顶升下垂梁架
The Usage of Lifting Jack

梁架抬升前后对比
Contrast of Beam frame Lifting between Start and Finish

木构件的油饰

木构件包括上架大木和下架大木。此次维修上架大木的梁枋、檩都有彩绘，椽梁部分油饰存在地仗开裂、褪色严重等问题，大部分圆椽及飞椽本身油饰保存较好，只是存在褪色问题。下架大木均为单色油饰，保存情况差别很大。在檐口之内的下架大木地仗基本上完好，在檐口之外的下架大木，地仗脱落严重，油饰保存状况较差。旗杆上所有地仗全部脱落。对于上架大木，重点放在檐口部位。按传统做法做单皮灰地仗三遍，然后刷醇酸漆三遍。完后做旧处理，避免过于鲜艳，与彩画不协调，局部修补，保持现有色调。对于只褪色而地仗完好的构件，只做除尘、刷漆处理。对于下架大木，没有地仗和地仗有裂缝、起皮的全部砸掉，表面处理干净，重新做一麻五灰地仗，刷单色漆，调配色彩，达到统一协调为目的。对于地仗很好的只做除尘刷漆处理。

Paint coat decoration on wooden components

The wooden components of ancient architectures include shelve wood and underbeam wood, the maintaining parts contains beam, purlin and colored paintings, some paint coat base layer on rafter cracks and fades severely, paint coat on most rafter and flying-rafter are preserved well which only fade a little. Underbeam woods which all painted by monochrome color show different preservation condition. The base layer on underbeam wood inside the cornice is basically fine while those outside are not good. The base layer on flagpole all drop down. For the shelve wood, the cornice should be stressed. The traditional way is to paint base layer with a kind of specific ash three times first and then alkyd paint three times too. Considering that the color of newly painting may be too bright, aging treatment is introduced to harmonize the new paint with old ones. For the wooden components which only fade while the base layer is intact, dedusting and repainting works are enough. As for underbeam wood, which without base layer or the layer is damaged, should all be polished and repainted with monochrome color to fit the unify standard. Dedusting and repainting works are only for the components which the base layer is intact without any problem.

檐口重做地仗
Repainting Work for Cornice Base Layer

檐口重新油饰
Repainting Work for Cornice Colored Drawings

贴金
Gold Foil Painting

制作地仗
Base Layer Manufacture

大门门槛油饰
Gate Threshold Painting

大门框架油饰
Gate Frame Painting

屋面的维修

本次维修的建筑屋面结构如下：

（1）木椽；

（2）望板；

（3）桦树皮；

（4）木垫板；

（5）木制筒瓦梗；

（6）铁皮瓦面。

经过检查，大门屋面和中牌楼屋面铁皮瓦锈蚀较严重，部分屋面漏水；顶部脊饰件用石膏制作，其余铁皮脊饰件有缺失破损。针对此情况，在屋面的处理时，在原有基础上，增加了一层油毡、铁皮防水层，在檐口加封铁皮条遮盖垫板头，防止日晒、雨淋；铁皮瓦全部采用一筒一板搭接，拆除的铁皮构件均做防锈处理。未拆除铁皮构件在屋面上除锈，然后调剂油漆色彩，接近现存的外观颜色。

为确保维修的观感效果，顶部石膏脊饰件更换为镀金脊饰件，缺失铁皮脊饰件和大门的猫头滴水均在国内加工制作。

Roof covering maintenance

The structure of roof covering maintained this time is as follows: 1. wooden rafter, 2. roof boarding, 3. birch-bark, 4. wooden underboarding, 5. arc-shaped tile, 6. iron sheet tile. The iron sheet tiles on the roof of front gate and middle archway corrode severely when checking, part of roof have the problem of leaking; ridge ornaments on the top are made of gypsum, others iron sheet decorations have been damaged. For that, linoleum and iron sheet waterproof layer are added on former foundation, iron straps can be seen at cornice to cover the head of underboarding to keep off sunlight and rain; iron sheet tiles all joint by tube and board, rust-proof works have been done on replaced iron components, for those iron components without replacement, derusting work is ok, after that, repaint with right color which is similar to the existing one.

To guarantee the visual effect of the maintenance, the gypsum ridge ornaments on the top should be replaced by gold-plated ones, ridge ornaments without iron sheet and cat-head driptiles should be made domestically.

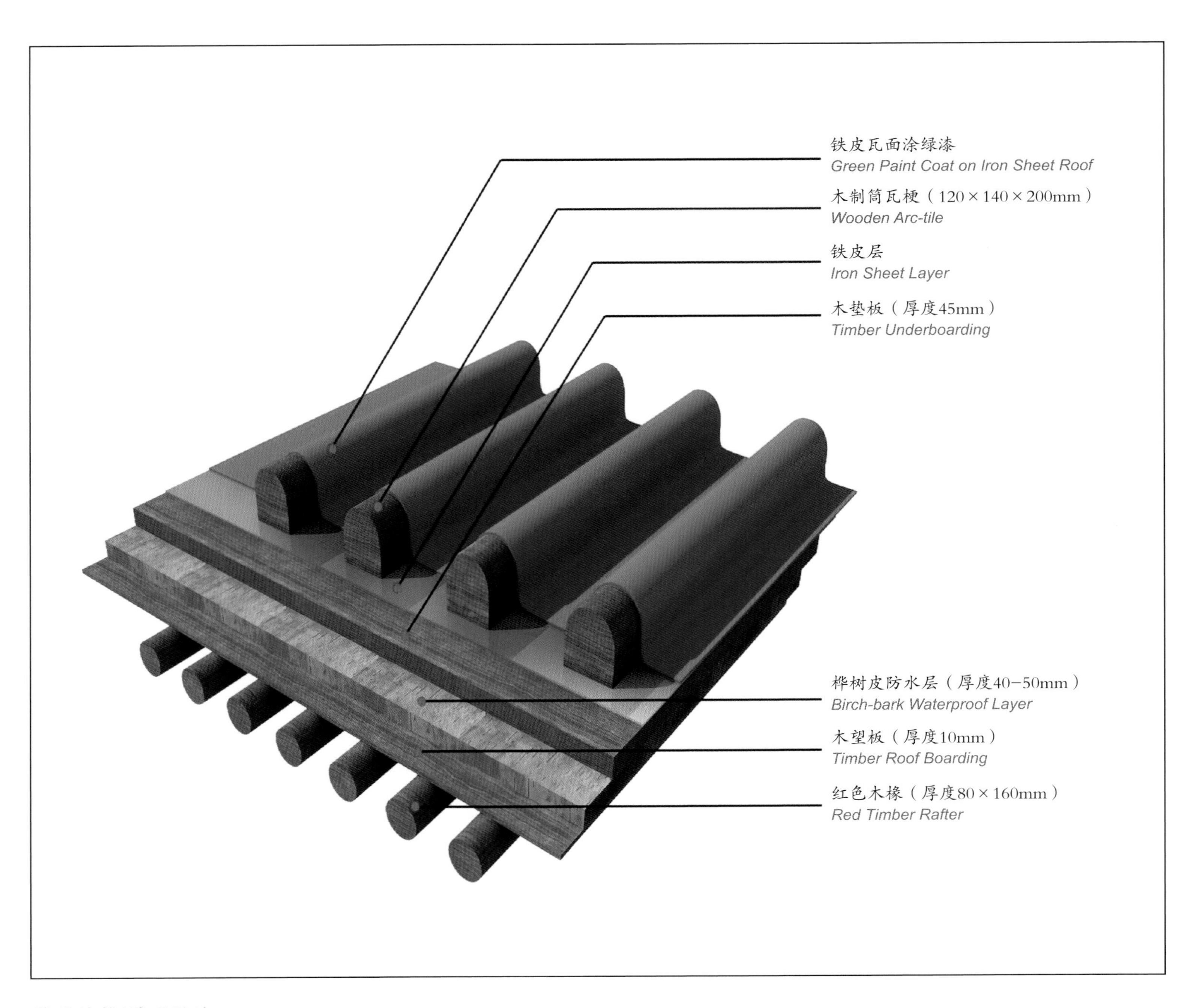

屋面结构/屋顶做法
Structure of Roof Covering

新作铁皮瓦件
Newly-made Iron Sheet Tiles

屋面破损
Roof Covering Damaged

石膏脊饰件
Gypsum Ridge Ornaments

搭接混乱
Disordered Connecting

铁皮瓦件锈蚀
Condition of Rotten Iron Sheet Tiles

铁皮脊饰件缺失
Absence of Iron Sheet Ridge Ornaments

大门铁皮脊饰件
Iron Sheet Ridge Ornaments at the Gate

大门铁皮脊饰件
Iron Sheet Ridge Ornaments at the Gate

铁皮瓦件安装
Erection of Iron Sheet Tiles

铁皮脊饰件维修
Maintenance of Iron Sheet Ridge Ornaments

铁皮佛像钉帽安装
Erection of Nail-head at Iron Sheet Buddha

屋面瓦件油漆
Roof Tiles Repainting

脊饰件的临时保护
Temporary Protection for Ridge Ornaments

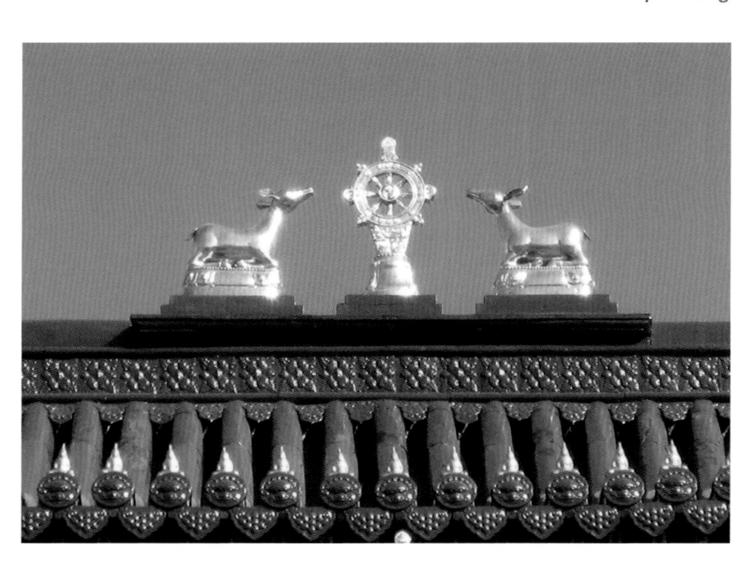

安装后的镀金脊饰件
Erected Gold-plated Ridge Ornaments

收藏在博物馆的鎏金脊饰件
Gold-plated Ridge Ornaments collected in Museum

在中国新制作的镀金脊饰件
Gold-plated Ridge Ornaments newly made in China

在中国新制作的镀金脊饰件
Gold-plated Ridge Ornaments newly made in China

大门外檐彩绘
Colored Painting on Exterior Gate Cornice

大门屋顶
Gate Roof

大门屋檐局部
Gate Eave

大门维修后局部
Maintained Gate

2.3.2 东、西便门维修

整体落架重修；重做基础，整体加固东、西便门及南宫墙基础；结合地面铺装抬升柱子，校正梁架。加固榫卯，添补缺失构件；增设台明及地面铺装，添补柱础，增设散水；油饰屋面及大木下架、门扇。

2.3.2 Maintenance of east-west side doors

Repair the whole components; rebuild the foundation, consolidate the east-west side doors and partial foundation of bounding wall; upraise the pillars according to the ground elevation, rectify the beam frame. Reinforce mortise and tenon joint, supplement absent components; pave platform and resurface the ground, repair the column base, build drainage platform; paint the roof, underbeam components and door leafs.

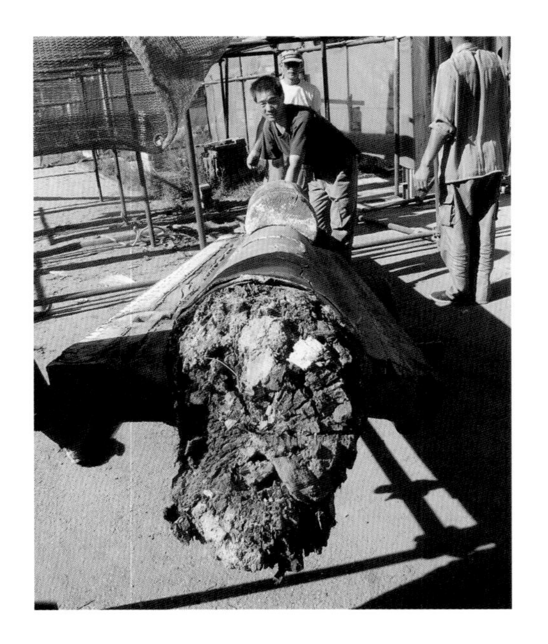

木柱腐朽状况
Condition of Decayed Pillar

木柱安装
Erection of Wooden Pillar

梁架安装
Erection of Beam Frame

铁皮防水层铺设
Iron Sheet Waterproof Layer Paving

屋面木梗安装
Erection of Roof Wooden Shaft

制作地仗
Base Layer Producing

博风板油饰
Paint Coat on Gable Eave Board

便门维修后
Repaired Side Door

南宫墙
South Palace Wall

木枋安装
Erection of Wooden Tiebeam

宫墙屋面安装
Erection of Palace Wall Roof

铺设铁皮防水层
Iron Sheet Waterproof Layer Paving

墙板安装
Wallboard Erection

渗桐油
Chinese Wood Oil Painting

表面油饰
Surface Painting

维修后宫墙
Condition after maintenance

2.3.3 牌楼维修内容

揭除屋面，校正梁架，加固榫卯。墩接糟朽柱子、戗柱。更换糟朽仔角梁，修补损坏花板；增设台明，重新铺装地面，更换混凝土夹杆石为石材夹杆石，增设散水；油饰大木下架；屋面除锈，重新油漆。

2.3.3 Maintenance of decorated archway

Remove the roof covering, correct the beam frame, strengthen mortise and tenon. Connect decayed pillars and props. Replace the decayed beam, repair the damaged board; repave the platform and ground, replace the concrete pole clamping plinth with stony one, pave drainage platform; paint the underbeam wooden components, clean the rust on roof covering and repaint it.

墩接木柱
Pillar Connecting

柱子纠偏
Pillar Deviation Correction

基础处理
Foundation Treatment

修复后
Plinth Condition after Maintenance

纠偏前
Condition before Deviation Correction

纠偏后
Condition after Deviation Correction

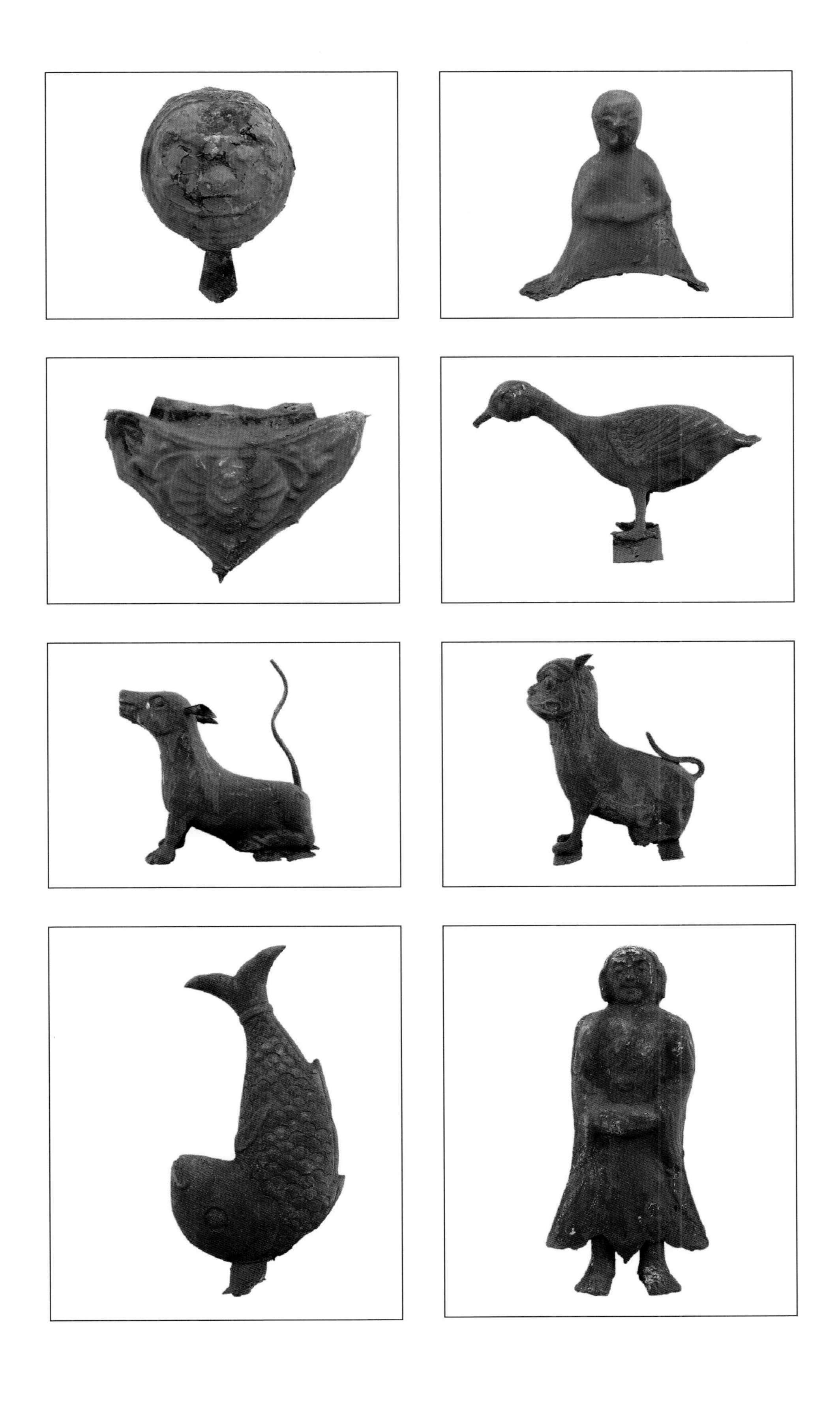

屋面小跑兽与铁皮瓦件
Decorations and Iron Sheet Tiles on the Roof

脊饰件安装
Erection of Ridge Ornaments

铁皮瓦件安装
Erection of Iron Sheet Tiles

脊饰件安装
Erection of Ridge Ornaments

屋面瓦件油饰
Roof Tiles Painting

屋面脊饰油饰
Ridge Ornament Painting

屋檐檐望油饰
Cornice Painting

维修后局部
Condition after Maintenance

维修后局部
Condition after Maintenance

柱子表面糟朽
Condition of Decayed Pillar

柱子糟朽
Condition of Decayed Pillar

巴掌榫墩接
Connection Work

木构件油饰
Wooden Component Painting

东西牌楼维修 ***East-West Archways***

维修后局部
Condition after Maintenance

2.3.4 旗杆的维修

2.3.4 Flagpole Maintenance Work

旗杆
Flagpole

搭架与支撑
Erection of Frame and Jackstay

柱子糟朽
Decayed Pillar

基础处理
Foundation Treatment

地面铺设
Ground Paving

地面及散水
Ground and Drainage Platform

基座加固处理
Plinth Strengthen work

维修中
Maintaining

维修后
Condition after Maintenance

2.3.5 栅栏墙的复原

2.3.5 Fence wall Recovery

复原后 *Condition after Recovery*

2.3.6 照壁的维修

照壁是本次维修建筑群唯一的砖结构，针对存在的问题主要进行以下内容的维修：（1）整体支撑加固；（2）基础圈梁埋植；（3）墙体整体校正；（4）南侧底座、池心更换；（5）墙体裂缝掏补；（6）瓦面刷隔水剂防渗处理；（7）散水重新铺设。

2.3.6 Maintenance of entrance screen wall

The entrance screen wall is the only brick structure architecture among groups of buildings, considering the existing problems, the maintenance work includes: 1. support and strengthen integrally; 2. embed basic ring beam; 3. rectify integral wall body; 4.replace south pedestal and center of the sculpture; 5. fill the wall fissure; 6.paint waterproofer on tiles; 7. repave the drainage platform.

修复前
Condition before Maintaining

墙体校正
Wall Body Rectify

整体支撑
Integral Support

圈梁埋植
Ring Beam Embedding

南侧底座更换
South Pedestal Replacing

维修后顶部
Tiles after Maintenance

维修后须弥座
Pedestal after Maintenance

维修后南面池心
South Center of Sculpture after Maintenance

维修后八字部分
"Eight" shaped section of screen wall after Maintenance

维修后
Condition after Maintenance

2.3.7 门前区广场的维修

施工内容为：各单体增设台明；统一调整广场地面标高，依原道路格局重新铺设路面。铺设停车场地面；重新设计广场排水走向，增设地下排水管网。路面铺设时，将原来的沥青路面全部更换为混凝土石子路面。

2.3.7 Maintenance of the front square

The maintenance work includes: pave platform; adjust the ground elevation, repave the ground according to the former ground condition. Pave the parking ground; redesign drainage trend at the square, build underground drainage network. Replace the former bituminous ground with concrete gravel pavement.

维修前路面

Blue Brick Ground before Maintenance

路面铺设
Ground Paving

表面打磨
Surface Polishing

广场地面铺设
Square Ground Paving

石子路面铺设
Gravel Road Paving

石子路面
Gravel Ground

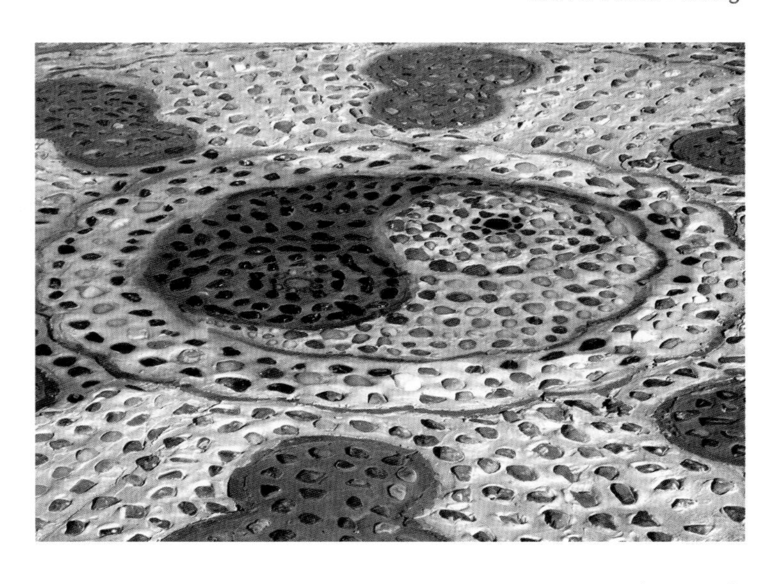

广场中心图案
Pattern at Square center

铺设后的广场地面
Square Ground after paving

3.彩画的保护修复

由于彩画(特别是外檐彩画)易受外部自然环境因素如温湿度周期变化、日晒雨漂、积尘污染等的影响而风化，一般只能保持50年左右的时间。如何能使残留至今的彩画保存下来，并对已有各类病害的彩画进行保护修复，是一个十分重要的课题。

彩画是东亚地区传统建筑文化特有的现象，文化多样性的体现。因此，彩画保护问题受到东亚各国的共同关注。2008年11月，在北京召开了“东亚地区木结构建筑彩画保护国际研讨会”，并通过了“关于东亚地区彩画保护的北京备忘录”。备忘录指出：“彩画保护的目的是最大限度地保留其真实性和完整性。”“彩画的保护应遵循科学的保护程序，以深入的研究和价值评估、现状评估以及管理评估为基础，确定保护方式。”“对彩画的干预应遵循最小干预、可逆性、可再处理以及干预技术和材料的适当性和兼容性等原则。”

西安文物保护修复中心经过多年研究和彩画保护修复实践，确立了彩画保护的前期调查、表面污染物清洗、加固、修补、封护的基本技术路线，并成功地应用于彩画保护修复的实践之中，如西安鼓楼、西安城墙长乐门彩画保护，天水伏羲庙彩画保护修复、河南嵩山少林寺彩画、重庆湖广会馆木构件雕饰贴金油饰等，为彩画保护理论和技术的发展奠定了基础，同时这些方法和技术也符合国际文物修复理论和原则，并完全体现于博格达汗宫建筑彩画的保护修复之中。

在博格达汗宫博物馆彩画保护修复中，首先对彩画保存现状进行了详细的调查纪录，包括照片纪录、文字纪录，并做全程录像纪录；将前期资料进行归类整理，完成图像编辑处理、病害现状的标识。通过剖面显微观察和显微红外光谱及X荧光等现代科学分析手段分析了博格达汗宫博物馆建筑彩画的制作材料和工艺特征，分析结论表明，博格达汗宫博物馆建筑油饰彩画工艺与中国彩画传统工艺基本相同。在大量科学研究和分析基础上，并通过保护修复前期试验，明确彩画清洗、加固、回贴和黏结、补全、表面封护等基本工艺和使用材料。

3.1 彩画的前期勘察和设计

彩画前期勘察是在现场对彩画的保存现状进行详细的测绘，现状记录、病害种类的统计，通过科学的检测分析，了解彩画的制作工艺及其所使用材料，并进行病因分析。还有根据病害类型，依据经验进行现场的保护修复试验，以确定采取何种保护修复技术措施，为保护修复方案的设计提供科学依据。

3.1.1 现状记录和病害勘察

现状记录：采用数字照相和现场临摹方式，对彩画的保存现状进行记录，真实地反映出彩画的现状和特征。

绘制彩画的病害图。彩画病害图是反映彩画各种病害的图示记录。通过病害图可以了解彩画病害的种类、面积，为方案设计和后期的维修保护提供基础的数据。

现场病害的微观形貌观察。通过对病害微观观察，可以深层次地了解病害状况和危害方式，为保护修复方案设计提供翔实的、有用的信息。采用便携视频显微镜，对各类病害进行微观形貌观察和分析。

3.1.2 彩画制作工艺和材料分析

了解彩画制作工艺和材料，是实施保护修复的前提。因此，在方案设计和保护修复前，必须对其制作工艺和材料进行深入的科学分析和研究。

博格达汗宫博物馆历经百余年的历史，但针对彩画工艺、材料及修复史几乎没有资料记载。为此，本项研究采取6个有代表性的彩画样品，通过剖面显微分析和显微红外、X荧光灯分析手段，对其制作工艺和材料进行了检测分析。

通过剖面显微观察分析及FTIR. XRF分析结果显示，博格达汗宫门前建筑大木构架均采用中国传统建筑工艺进行彩画，立柱、梁枋大件的地仗多为一麻五灰，小件及花窗枋板则为单皮灰地仗，其披麻工艺规范到位，做工考究。桐油用量相对较足，披麻性能较好。比如大门建筑，大木构虽历经百余年风雨，但地仗层较少有空鼓、开裂、脱落之病变。 从彩画地仗的剖面观察发现，大门1-4檐及东西南北四面地仗层均为单皮灰。腻子为中国传统配比，即猪血、砖粉、桐油。彩画层基本工序为：拍谱—沥粉—填色（大色、二色）—贴金—挂粉（白色）—压老（黑色）等步骤。比如，梁上：单皮灰—作底色(底色不同)—施彩；蔓草图案：单皮灰—红底色—沥粉—施彩—贴金；人物风景画：单皮灰—白色打底—绘画；枋上和梁上祥云：单皮灰—直接施各种彩画。

彩画用色特点：彩画主要色调以金、蓝、绿、红为主，其次还有白、黑、黄等色作点缀；其次是施金量大，在梁、枋、花板、麻叶头等部位都大量采用了贴金和沥粉贴金工艺，彩画等级较高；另外枋、花板和抱厦天花用色对比强烈，而梁上及走马板上色彩淡雅，手法和工艺形成鲜明对比，是博格达汗宫彩画工艺的又一特点。

通过剖面显微观察和显微红外光谱及X荧光等分析手段对蒙古国博格达汗宫建筑彩画的制作材料和工艺特征的分析，剖面显微观察显示历次彩画重彩层次结构；紫外显微剖面观察还发现了彩画层中施胶的信息。分析结果还显示：博格达汗宫建筑彩画工艺与中国彩画传统工艺基本相同。

3. Conservation and restoration of the colored drawings

Colored drawings(especially exterior eaves) are vulnerable to external natural environmental factors such as the cycle changes of temperature and humidity, sun and rain, dust pollution, etc. and can be easily air-slacked, they can only last for about 50 years. How to preserve the residual colored drawings and repair the damaged parts, this is a very important issue.

Colored drawing is a unique phenomenon in traditional architectural culture in East Asian region, it is the embodiment of cultural diversity. Therefore, the issue of colored drawing protection is a common concern of East Asian countries. In Nov. 2008, "International Seminar on East Asian Wooden Structure Colored Drawing Conservation" was held in Beijing, and "Beijing Memorandum" was passed. The memorandum stated: "The purpose of colored drawing conservation is to maximize the of its authenticity and integrity." "Scientific conservation procedures should be followed, in-depth research, valuation, situation assessment and management assessment should be the base to ensure the protection." "The principles of minimum intervention, reversibility, reprocessing, appropriateness and compatibility of the materials should come into notice."

After years of study and colored drawing protection practice, Xi'an Centre for the Conservation and Restoration of Cultural Heritage conducted the pre-investigation of colored drawing protection and designed the basic technical route of surface contaminant cleaning, reinforcement, repairing, protective coating, and successfully applied them to colored painting conservation and restoration practice, such as paintings on Xi'an Drum Tower, Changle Gate of Xi'an City Wall, Fuxi temple in Tianshui, Shaolin temple in Henan Mount Song, wood components carving gold oil ornaments in Huguang Assembly Hall in Chongqing, etc. which laid a foundation for the theories and technologies of painting protection, these approaches and techniques are also compliance with theories and principles of international cultural heritage conservation, and are fully reflected in the maintenance project of Bogd Khaan Palace Museum.

During the process of colored painting conservation and restoration of Bogd Khaan Palace Museum, a detail investigation records were first conducted over the conservation situation of colored paintings, including the photo records, text records and video records; the preliminary data was classified and organized, image editing and processing were completed, disease condition was also marked. Modern scientific analysis approached such as microscopic observation, micro-infrared spectroscope and X-ray fluorescence were adopted to analyze materials and processing features of colored paintings of Bogd Khaan Palace Museum, the analysis concluded that the oil painting crafts of Bogd Khaan Palace

Museum were basically the same as traditional Chinese colored painting process. On the basis of a large number of scientific research and analysis, along with preliminary tests, the basic process and materials used in painting cleaning, reinforcement, bonding, repainting, protective coating etc. were clear.

3.1 Preliminary investigation and design over colored paintings

Preliminary investigation over colored paintings is to conduct detail field mapping for the conservation state of colored paintings, take records for conditions and count the disease types, understand the production process and materials and analyze the cause of disease through scientific testing and analyzing. Determine the technical measures taken for colored paintings protection to provide a scientific basis to conservation program according to the type of disease and based on experience conducted on-site.

3.1.1 Status records and disease survey

Status records: The ways of digital photography and field copying were adopted to record the preservation status of colored paintings to truly reflect the status and characteristics of the colored drawings.

Draw the figure of disease of colored paintings. The figure of disease of colored paintings is a record to reflect the variety of diseases of colored paintings, the figure can show the type of the disease and the size, it can provide the basis for program design and post-maintenance protection.

Observe the micro-image of the field disease. Disease condition and harming way can be deeply understood by observing the micro-image of the field disease, which is to provide detail and useful information for conservation and restoration program design. Portable video microscope is adopted to observe and analyze the image of all kinds of diseases.

3.1.2 Colored drawing production process and materials analysis

The premise of conducting the protection program is to understand the production process and materials. Therefore, before designing the program and practicing it, in-depth scientific analysis and research over production process and materials should be introduced.

Over hundreds years of history, there are almost no data for colored painting techniques, materials and repairing history of Bogd Khaan Palace Museum. For this, the study adopt six typical samples, modern scientific analysis approached such as microscopic observation, micro-infrared spectroscope and X-ray fluorescence were introduced to analyze materials and processing features of colored paintings.

Through microscope observation and analysis of FTIR. XRF, the wooden components of buildings in front of Bogd Khaan Palace Museum all show Chinese traditional architectural process on colored paintings, the base layer on vertical columns and roof beams largely use a kind of Chinese traditional method, the process is compliance with specifications with exquisite workmanship. The usage amount of Chinese wood oil is respectively sufficient. The base layer on big wooden components of the front gate seldom crack and drop. Through observation of the base layer of paintings, the eaves from the first to the fourth storey and base layer on east, west, south and north directions are all one ash layer. The putty is Chinese traditional mixture, including pig blood, brick powder and Chinese wood oil. The basic processes of the paintings are all strict and totally complex.

The color features: the main colors of the paintings are gold, blue, green and red, followed by white, black, yellow and other colors as a decoration; the amount of gold powder is huge, especially at beam, square column, board, Mayetou, the rank of the colored paintings is high; additionally, there are strong color contrast at beam, board, baoxia and ceiling, while the color at roof beam and zouma board is light and elegant, the techniques and processes show a strong contrast, which is another feature of painting process.

By analytical means such as profile microscopic observation, microscopic infrared spectrum and X fluorescent light, we analyzed the materials and craft characteristics of architectural colored paintings of Bogd Khaan Palace Museum, the profile microscopic observation showed the structure of former repainting, UV-light microscopic observation revealed the existence of glue at painting tier. Analytical conclusion: the colored painting craft from Bogd Khaan Palace Museum is basically the same as Chinese traditional colored painting craft.

颜料层脱落
Color Coat drops

鸟窝及鸟粪
Bird's Nest and Droppings

地仗脱落
Base Layer drops

龟裂
Crack

水渍
Water Spots

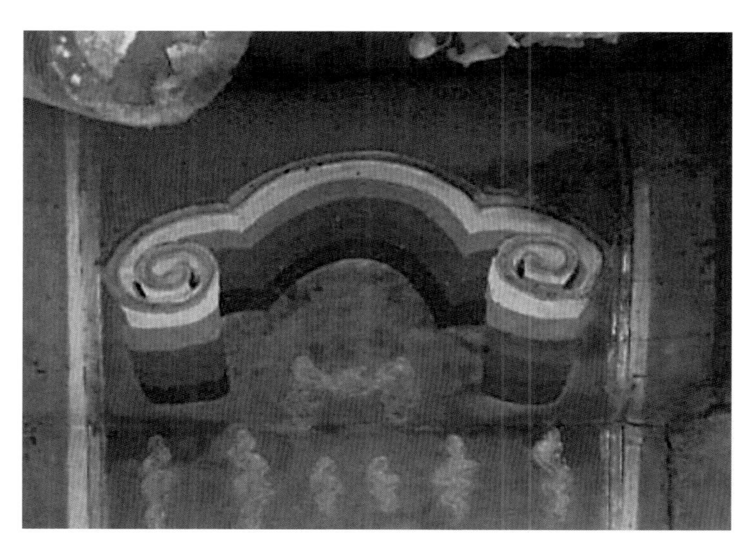

油污
Greasy Dirt

彩绘开裂脱落
Colored Painting cracks and drops

裂缝
Fissure

木构件基层、地仗起翘开裂
Base and Base Layer of Wooden Structure peel and crack

局部木构件脱落、缺损
Part of Wooden Structure dropped and damaged

现场显微照相
Field Microphotograph

苔藓地衣
Moss Lichen

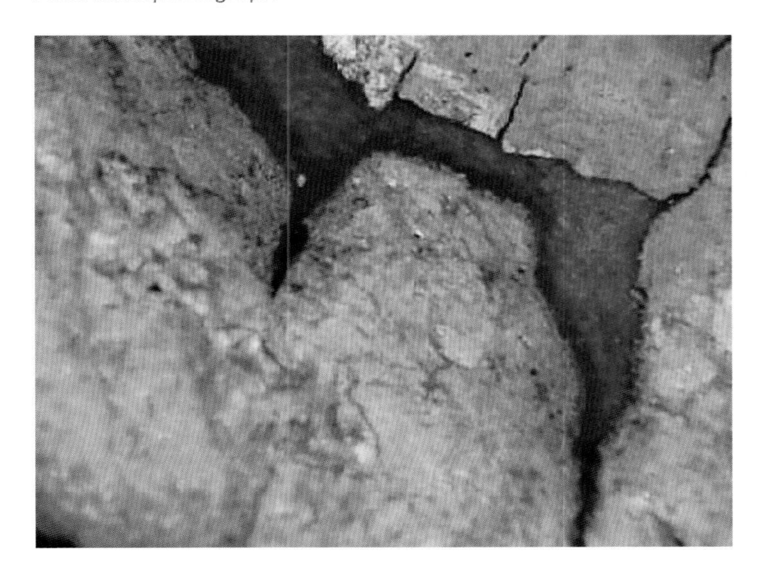

金层起翘
Golden layer peeling

金油泛出
Golden Oil Overflowing

单皮灰剖面
Lime Profile

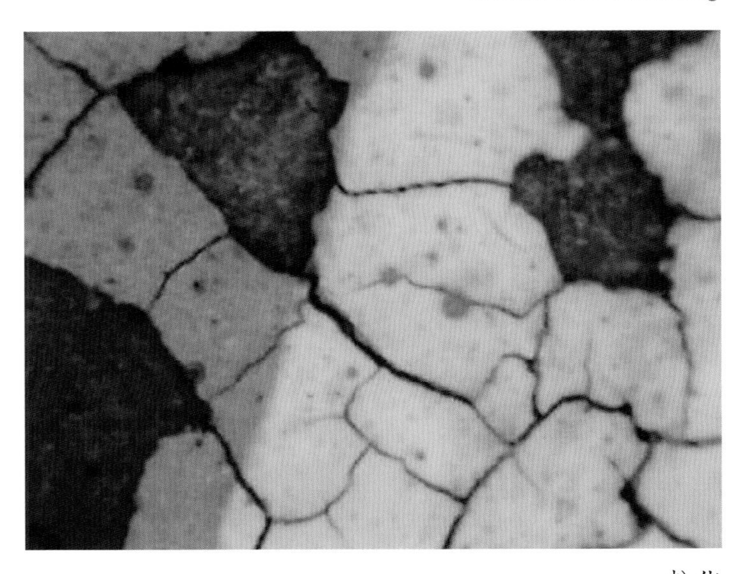

粉化
Pulverization

照片拼图
Photo Collage

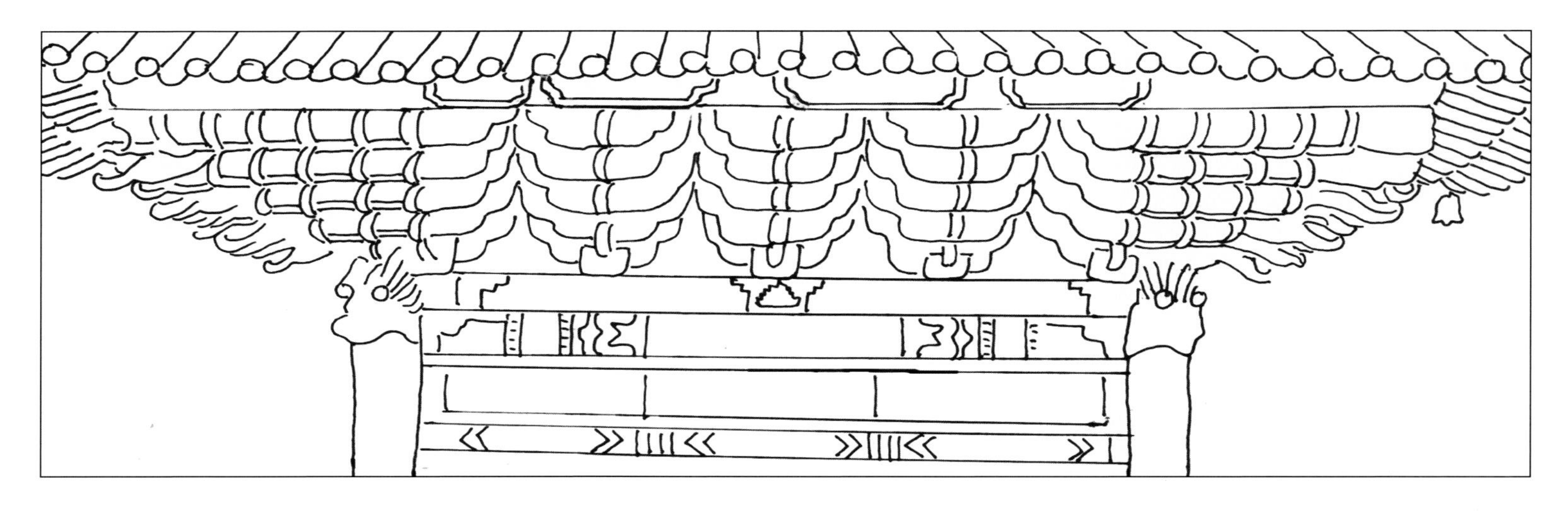

线图
Line Graph

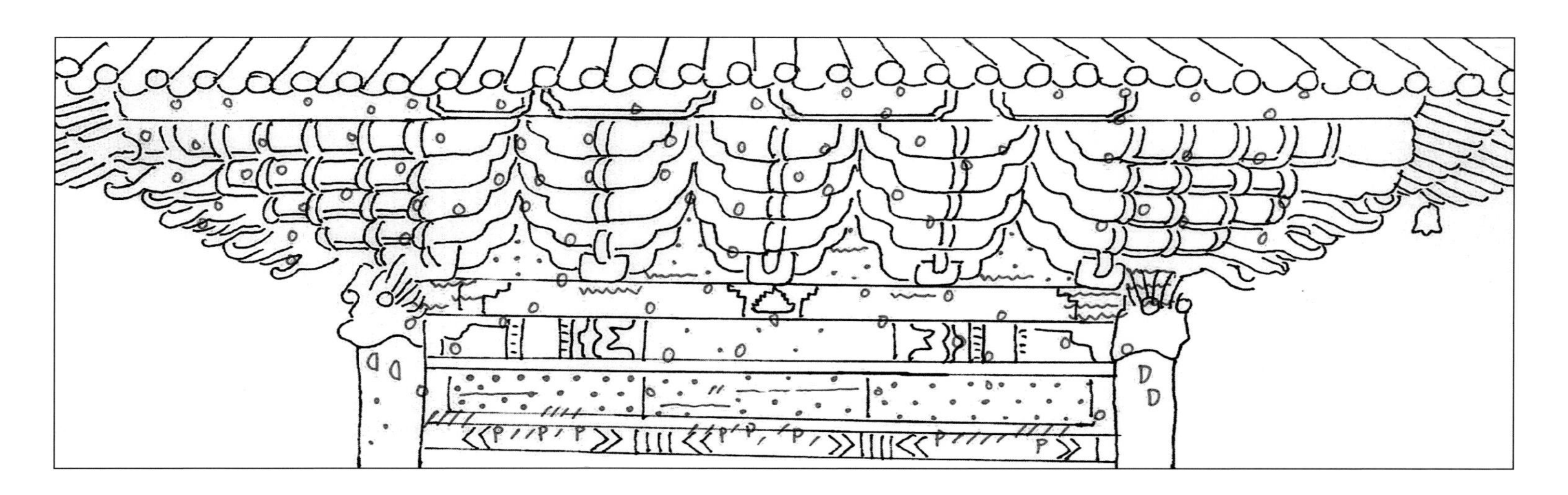

病害图
Photo of disease

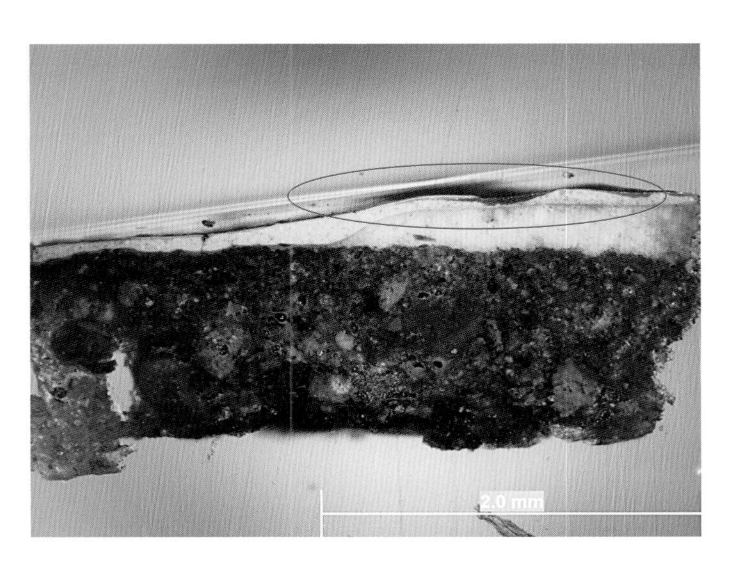

图 1: MG-1. (Vis.50x)
Picture 1: MG-1. (Vis.50x)

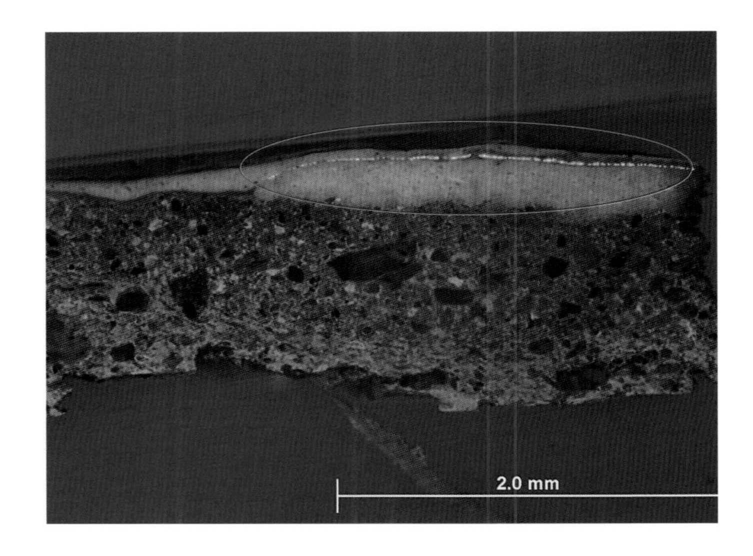

图2: MG-1. (UV.50x)
Picture 2: MG-1. (UV.50x)

图3: MG-6. (VIS100x)
Picture 3: MG-6. (VIS100x)

图4:MG-6 (UV.100x)
Picture 4:MG-6 (UV.100x)

图5: MG-2 (vis.50x)
Picture 5: MG-2 (vis.50x)

图6: FTIR

①MG-6 红色箭头点红外谱图 ②熟桐油红外谱图

MG-6 red arrowhead infrared spectrogram *Infrared spectrogram of China wood oil*

样品MG-1在可见光下彩画层的剖面显微分析显示为多层，从颜料层到地仗层层次依次为白色层—金层—暗绿色层—白色层—黄色层—蓝色层—绿色层—白色层—地仗层，当在紫外光下该样品在沥粉层之上出现了一条较强的黄绿色荧光(图1、图2) 。

样品MG-6的剖面显微照片显示，在可见光下彩画层的层次依次为红色层—橙色层—白色层，也就是说该样品反映出油饰彩画层的基本制作程序是三道，但当在紫外光下该样品剖面层次分成了6层(图3、图4)，图3中的红色层在图4中分成了4层：白色层—红色层—暗红色层-白色层。紫外光下红色层显示出4层应该是不同时期重彩形成的，其中铅丹层之上的白色层应为胶层，胶层之上的暗红色层为早期油饰彩画层，暗红色层之上的亮红色层为后期重彩层，而亮红色层之上的白色层应为罩的桐油层。在紫外光下进一步揭示出了可见光下未能反映出的彩画层工艺细节及程序。

MG-2是在地仗层上先施彩（图5箭头所指绿色层）后沥粉贴金，同时在该地仗层下还发现一彩画层（图5 红色圈线内）。

显微红外光谱及X荧光分析结果表明，地仗层表面的白色层检测出有机物和方解石粉，第二层橙黄色为铅丹，第三、四、五层的红外谱图基本一致，有桐油的特征峰，同时X荧光分析检测出了Pb、Ca元素，在第六层同样检测出桐油（图6）。

Profile of colored drawing layer of sample MG-1 shows multilayer under microscopic analysis, the color from pigment layer to base layer is successively white-golden-deep green-white-yellow-blue-green-white-base layer, there appears yellow green fluorescence above gold powder layer under UV-light. (Picture 1, Picture 2)

The micrograph of the profile of sample MG-6 shows that the colored painting layer under visible light are successively red-orange-white, which reveals the three steps of making colored drawings. While under UV-light, this profile of sample MG-6 turns to be six layers (Picture 3 and Picture 4), the red layer in Picture 3 separates into four layers in Picture 4, which is white-red-deep red-white. Those four layer under UV-light indicate the repainting works during different times, the white layer over miniumite layer should be adhesive layer, the deep red layer over the miniumite layer is colored drawing from early stage, while the bright red layer over the deep red layer is repainting layer, and the white layer over the bright red layer should be Chinese wood oil. The UV-light can further reveal the details and processes of the colored drawings which can not be seen under visible light.

The production process of sample MG-2 is first colored painting (the green layer marked in Picture 5) and then gold-plating on Base layer, a layer of colored painting is also found under base layer. (Marked inside the red circle in Picture 5)

By using microscopic infrared spectrum and X fluorescent light analysis methods, the white layer, which is of the surface of the base layer, is detected containing organic matters and calcite powder, the second yellow layer is miniumite, the infrared spectrum of the third, fourth and fifth layer are generally consistent with Chinese wood oil, elements of Pb and Ca are also found by X fluorescent light analysis, Chinese wood oil exists in the sixth layer too. (Picture 6)

3.1.3 现场试验

清洗方法筛选试验

分别采用去离子水、有机溶剂和水混合溶液，以及有机溶剂进行了现场的清洗试验。实验结果表明，有机溶剂对彩画没有影响，适合使用。对于风化比较严重的贴金部位，金层容易被擦除，对贴金造成损害，对此类型贴金部位采用水溶性的试剂进行清洗。

3.1.3 Field Test

Screening experiment on cleaning method

We respectively use deionized water, organic solvent and Water mixture. Test result indicates that organic solvent has no adverse effect on colored drawings, but the gold foil layer can be easily wiped off. Therefore, the proper method is to use water-soluble reagent for gold-plated section cleaning.

序号	清洗溶液	实验部位	脱色评估	效果评估	备注
1	去离子水	边框	脱色	差	
2	水+乙醇	边框	脱色	差	
3	乙醇+丙酮	边框	微脱	一般	
4	丙酮+三氯乙烷	贴金部位	不脱	良好	
5	水+乙醇+氨水	烟熏部位		一般	

No.	Cleaning solution	Test part	Decoloration assessment	Effect assessment	Note
1	deionized water	Frame	Decolorate	Bad	
2	Water+alcohol	Frame	Decolorate	Bad	
3	Alcohol+acetone	Frame	Slightly decolorate	Normal	
4	acetone+ trichloroethane	Gold foil paint	Color stay	Fine	
5	Water+alcohol+ ammonia water	Fumigation part		Normal	

a）加固及封护材料筛选试验

对四种材料进行了现场试验，实验结果表明，采用2%的PARALOID B72的丙酮及三氯乙烷溶液进行加固和封护效果较好。

a) Screening experiment on protective coating material

After testing four materials, 2% PARALOID B72 acetone and trichloroethane solution should be used for protective coating.

加固实验材料及效果对比表

序号	加固剂名称	溶　剂	浓　度	加固效果	备注
1	聚乙烯醇缩丁醛	乙醇	1%　3%	1%效果差 3%表面泛白	
2	乙基纤维纤维素	乙醇	2%　3%	加固效果较差	
3	Paraloid B 72	乙醇	2%　5%	加固效果好	
4	Paraloid B 72	三氯乙烷	2%	加固效果较好	
5	Paraloid B 72	乙酸乙酯	2%	加固效果较好	
6	Remmers 300	乙醇	50%	提色效果好	

Consolidating test material and effect comparison table

No.	Fixation accelerator	Solution	concentration	Effect	Note
1	Polyvinyl butyral	alcohol	1% 3%	1% bad,3%surface whiting	
2	Ethylcellulose	alcohol	2% 3%	Slightly bad	
3	Paraloid B 72	alcohol	2% 5%	well	
4	Paraloid B 72	trichloroethane	2%	better	
5	Paraloid B 72	ethyl acetate	2%	better	
6	Remmers 300	alcohol	50%	well	

b）地仗主要出现的问题是开裂、起翘、局部脱落的补全处理

按照原制作工艺使用猪血腻子做地仗补全的材料。

b) The main problems happened on base layer are cracking, peeling and partial dropping. According to the traditional craftsmanship, pig blood putty can be painted on for repainting work.

c)特殊问题：褪色严重模糊不清的彩画复原的研究

显影——红外摄影技术的应用

由于现场有部分彩画图案已经缺失或者模糊不清，采用红外照相技术，开展彩画图案显现研究。

c) Special problem: Research on colored paintings which fade severely.

Developing method——Application of infrared photography technology

Due to the problems that some colored patterns are absent and unclear, on the basis of the current condition, we tried infrared photography technology with the hope of recovering some pictures.

红外灯加热
Infrared Lamp Heating

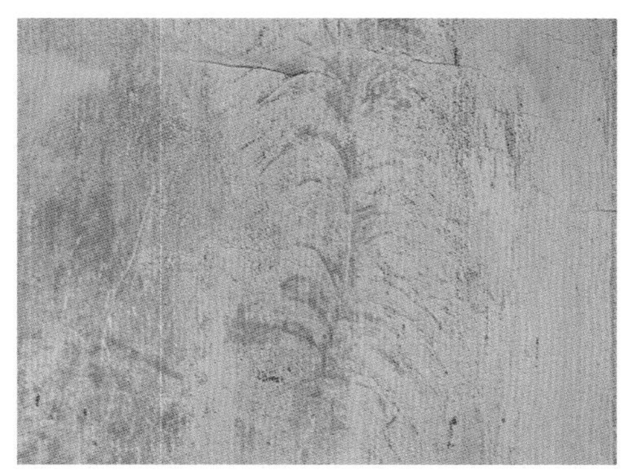

加热后的照片局部
Part of the Picture after Heating

未加热的照片
Picture before Heating

显影——传统材料及工艺的采用

东西牌楼及中牌楼梁枋局部彩画保存现状很差，地仗起翘、脱落，质地疏松，表面图案已经褪色，基本看不清楚，作为古建筑彩画已经失去了其重要的特性，即装饰性；地仗酥松，易脱落，失去了其对木基底的保护作用。由于该部分彩画需进行重绘，因此，采用传统材料——桐油混合溶液涂刷在表面，显影后将原图案复制下来做摹本，重做地仗后将原图案重新绘制在原处。

4.2 Developing method—Apply of traditional materials and crafts

Colored drawings on painted beam of east-west archways and middle archway are preserved awfully bad, the base layer peel and drop, patterns on the surface fade and are unclear, which have lost the decorating function; the base layer is loose and can easily drop without protective action. Therefore, the colored patterns need repainting, China wood oil, a kind of traditional material, should be paved on the surface, then copy the former patterns, repaint the patterns after the base layer renewed.

保存原状

Preserve the Original State

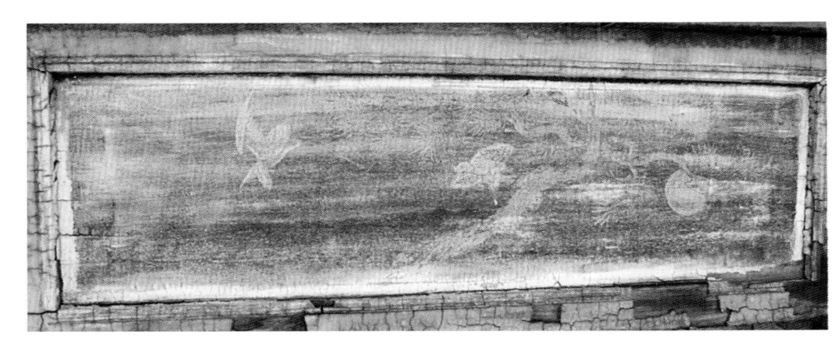

显影后的状况
Condition after using Developing Method

根据显影后恢复的图案
Recovered Pattern after using Developing Method

显影前
Before Developing

显影后
After Developing

3.2 彩画的保护修复

彩画的保护修复技术难点

较薄贴金部位的清洗。由于这部分贴金层比较薄，大多表面失去光泽，贴金的部位在彩画中起着重要的装饰作用，如果清洗不当，使贴金损失，会对彩画的整体视觉效果产生不良的影响，但是如果清洗不彻底，金的光泽没有清洗出来，同样会影响彩画的视觉效果。因此对贴金部位选择适合进行大面积清洗的材料及清洗的效果是非常重要的。

枋心画面的裂缝以及龟裂起翘起甲的现象十分影响画面的效果，裂缝及龟裂、起甲也不利于彩画的长期保存，因此，这个问题在保护修复中是必须处理的。但是由于画面均为小写意或是兼工带写的人物、花鸟及山水等内容，绘制细腻，裂缝及起甲纹理细小，补地仗有一定的难度，补绘的图案必须进行程度合适的做旧处理，并且补绘应符合原绘画的意境及韵味。应该注意的是地仗及补绘都不能覆盖现存的画面。

现场具体实施

主要有表面清洗，地仗起翘回贴，做旧补绘（地仗补全，表面颜色及图案的补全），表面封护。

3.2 Protection and maintenance of the colored drawings

Technical Difficulties：

The gold foil painting part which plays significant decorating function in colored drawings should be cleaned. Gold-plated layer is thin with dull surface, if it is cleaned inappropriately, the overall visual effect of colored drawing would be adversely affected. Therefore, the proper cleaning materials must be carefully chose.

The fissures and crack phenomenon on central portion of painted beam severely affects the drawing appearance which also goes against long-term conservation for colored paintings, so, this matter should come into notice in the process of protection and maintenance. However, because the drawing is a kind of small freehand brushwork in traditional Chinese painting containing people, birds, flowers and landscape which draws exquisitely, fissure and crack texture is subtle, base layer repainting work may have certain difficulties. Aging treatment should be done for newly painted patterns to make them look as the same as former ones, the repainting drawings should fit former artistic conception.

According to the preservation condition of colored drawings and earlier stage research, the specific maintaining process of field operation should be made, which contains earlier days test, surface cleaning, damaged base layer repainting, colored pattern redrawing (supplement work over base layer and surface colored drawing) and surface protective coating.

表面清洗

在前期实验的基础之上，使用丙酮、乙醇等挥发性的有机溶剂对彩画表面进行清洗，溶剂内不添加去离子水，并配合棉签、手术刀等工具。彩画表面没有油污等污染物。

Surface cleaning

On the basis of earlier stage test, the surface of colored drawings is not contaminated by greasy dirt, volatile organic solvent such as acetone and ethanol without deionized water can be used to clean the drawing surface. Together with tools such as cotton swab and bistoury, can be used to extensively clean the colored drawings.

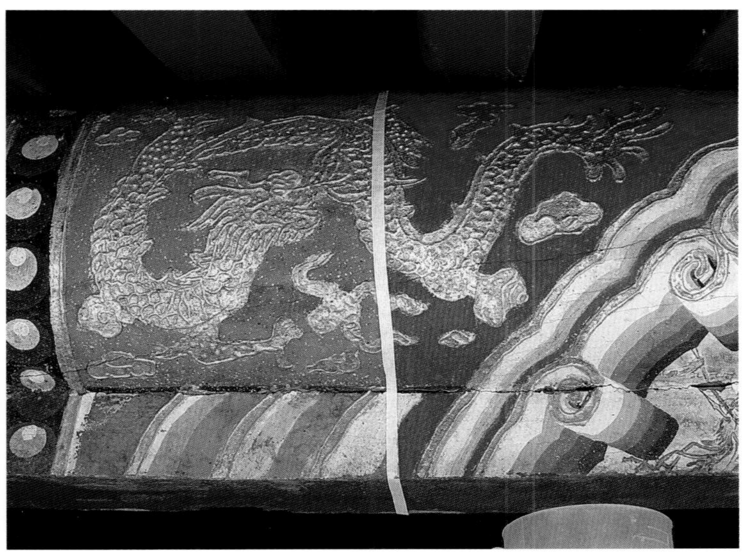

清洗效果对比照片
Comparison photos of cleaning effect

起翘回贴

彩画的局部有不同程度的起翘、起甲现象，使用羧甲基纤维素溶解成适合的浓度，用细针管吸取纤维素注射至地仗层底部，待地仗回软之后，用棉托按压回贴至原位，表面用压舌板固定。

Peeling part repainting

Due to the peeling phenomenon on colored paintings, carboxymethylcellulose could firstly be dissolved to proper concentration, then use fine needle tube to suck up the cellulose and inject to the bottom of the base layer, secondly press the injected place after the base layer turning soft, finally, fix it with spatula.

做旧补绘

补绘包括两方面：地仗的补全、表面颜色及图案的补全。

地仗的补全

按照原制作工艺使用猪血腻子进行补全，待干后打磨平整。针对枋心的裂缝及起甲的补全不能影响到画面的观赏性及完整性要求，使用了石膏及猪血腻子分别进行补全。

表面颜色及图案的补全

原彩画局部有脱落，地仗补全之后表面缺色，为整体图案的完整性及可观赏性，需要对颜色及图案的缺失部位儿行补绘。

参照原保存的图案及颜色进行做旧补绘，根据不同部位的颜料使用工艺及方法，将矿物颜料调制好，进行补绘，这是针对单色及没有绘画内容的部位，画面的局部补绘要根据绘画的内容、技法、神韵进行补绘。

Aging treatment and repainting Repainting work includes: base layer supplement; surface color and pattern supplement

Base layer supplement

According to the traditional craftsmanship, pig blood putty can be painted on for repainting work, then polish smooth when dry. The supplement work for fissures and peeling part on central portion of painted beam should not adversely affect the appreciation value and integrity of the drawings. Considering the effect and former craftsmanship, gypsum and pig blood putty are respectively used for maintenance.

Surface color and patterns repainting

The former colored paintings drop down, the surface appearance of base layer is color defective, which adversely affect the integrity and appreciation value of the overall pattern.

For problems above, repainting work and aging treatment should be done according to former patterns and color. Referring to different methods of pigment usage, concoct the mineral pigment for repainting which aims at the part where it is homophony or without drawings, partial repainting work should be compliance with the content, techniques and romantic charm of the colored drawings.

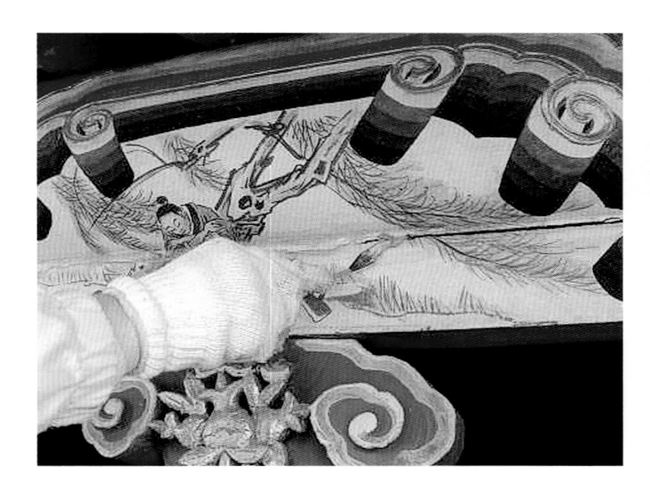

地仗补全
Base layer repainting

表面局部补绘
Partial repainting on surface

表面加固和封护

使用2%的PARALOID B72 的丙酮及三氯乙烷溶液进行加固和封护。

Surface protective coating

Use 2% PARALOID B72 acetone and trichloroethane solution for protective coating.

表面封护
Surface Protective Coating

修复前、中、后的照片效果对比

Photo Effect Contrast of pre-maintain, in-maintain and post-maintain

清洗前
Condition before Cleaning

清洗后
Condition after Cleaning

地仗补全
Putty Repainting

修复后
Condition after Maintaining

修复后
Condition after Maintaining

修复前
Condition before Cleaning

修复后
Condition after Cleaning

修复前
Condition before Cleaning

修复前
Condition before Maintaining

修复前
Condition before Maintaining

清洗后
Condition after Cleaning

清洗后
Condition after Cleaning

修复后
Condition after Maintaining

修复后
Condition after Maintaining

修复后
Condition after Maintaining

修复前
Condition before Maintaining

修复后
Condition after Maintaining

修复前
Condition before Maintaining

修复前
Condition before Maintaining

修复后
Condition after Maintaining

修复前
Condition before Maintaining

清洗后
Condition after Cleaning

修复后
Condition after Maintaining

修复后
Condition after Maintaining

修复前
Condition before Maintaining

修复后
Condition after Maintaining

修复前
Condition before Maintaining

修复前
Condition before Maintaining

修复后
Condition after Maintaining

4. 维修效果

2007年9月28日，经过中蒙文物专家组对维修工程进行的联合验收，认为该工程严格遵循国际公认的文物保护修复原则，所采取的技术措施合理，彩画保护修复依据可靠，较好地保存了文化遗产的真实性和完整性，工程质量和效果良好，一致通过验收。

4. Maintenance Effect

On Sep. 28th 2007, Chinese and Mongolian experts regarded this project as having strictly followed the international recognized principles of cultural heritage protection and maintenance after their acceptance check, the technology measures being taken were appropriate, the practice of colored paintings protection and maintenance was perfect which nicely preserved the authenticity and integrity of cultural heritages, the construction quality and effect were quite well which got unanimous approval.

肆

FOUR

The Maintenance Project of Bogd Khaan Palace Museum

博格达汗宫博物馆维修工程

其他

OTHER CONTENTS

其他

OTHER CONTENTS

The Maintenance Project of Bogd Khaan Palace Museum

博格达汗宫博物馆维修工程

4 其他

1.交流与培训

中国驻蒙两任大使高树茂大使和余洪耀大使，多次来到工地检查，并要求这个工程在蒙古国做出一个典范，以推广中国文物修复的理念，在工程实施中通过开展交流，展示我们中国人维护文物的高超技能。蒙古国政府官员、相关专家也多次到现场进行监督检查，并高度评价了维修的效果。美国、德国、瑞士等多国专家数次来到现场，与我技术人员进行交流，对维修技术表示钦佩。与此同时，工程先后召开了两次新闻发布会，共有中国和蒙古国10余家媒体到现场进行了采访和专访，蒙古国博格达汗宫博物馆馆长在发布会上，通过与后面维修的对比，对我们双方的合作十分满意，对维修组织和维修手段高度赞扬，认为维修的效果非常好。

在两年维修过程中，我单位先后派出30多人赴蒙开展科学研究与技术指导工作。期间，每个人都能严格遵守国家外事纪律及蒙古国法律法规，为援外工程树形象。没有发生一起违反外事纪律及当地法律的事情，受到驻蒙大使馆和合作方的高度好评，为中国文物保护树立了良好的形象。同时蒙方也先后派出10人次在现场进行学习实践与交流，学习古建筑和彩绘保护的理念，观摩维修方法，逐步实践维修的各个步骤。

2007年11月至2008年1月我们邀请蒙方4名人员到我单位进行学习交流和培训。在培训过程中，他们系统地学习了古建筑保护基本知识、古建筑维修技术规范、壁画彩绘保护等课程，进行了文物修复，最后取得了培训证书。通过培训，基本掌握了古建筑维修理念、方法和技术，对于博格达汗宫博物馆门前区的后续保护工作起到了重要作用。

2011年5月22日至5月24日，蒙古国文化遗产保护中心主任G.Enkhbat一行5人，来陕西省文物保护研究院考察访问。访问期间，双方进行了座谈，G.Enkhbat主任表达了此行目的：一是考察中国相关文物保护单位的机构、人员以及设备的情况；二是寻求合作，希望和陕西省文物保护研究院在寺庙彩画、壁画修复等方面开展合作。之后实地考察了陕西省文物保护研究院实验室、修复室，以及壁画保护修复现场。

1\Technology exchange and training

Gao Shumao and Yu Hongyao, two of Chinese ambassadors in Mongolia, came to inspect the construction site repeatedly and asked for the project to be a model in Mongolia so as to popularize China' s ideas in cultural heritage conservation and restoration and to display our Chinese superb skills in maintenance of cultural objects through technology exchange in project implementation. By supervising and inspecting the construction site, Mongolian government officials and experts highly praised the effect of maintenance. Experts from America, Germany, Switzerland, etc. also expressed their admirations for our maintenance technology after communication with our technical staff. Meanwhile, we successively held two press conferences for the project, more than ten media from China and Mongolia reported or exclusively interviewed on the spot. Compared with latter maintenance, curator of Mongolia Bogd Khaan Palace Museum extended his satisfaction on Sino-Mongolia cooperation in press conference. Moreover, museum curator also highly praised the maintenance organizations and means, thinking that the effect of maintenance was beyond compare.
In two years maintenance process, we successively sent more than 30 staff to Mongolia for scientific research and technical guidance. Every one of them kept strictly to the National foreign affairs discipline and laws and regulations of Mongolia during that period, contributing to set a positive image for foreign aid project. Furthermore, no event such as violation of foreign affairs discipline and local laws happened, which received great praise from the Embassy in Mongolia and cooperation partner, establishing a good image for Chinese cultural heritage protection. Besides, to the benefit of gradually practicing the maintenance of various steps, 10 people was sent successively by Mongolian side to project spot with the aim of practice learning and technology exchanging, including ideas of historical building, colored drawing or patterns protection and maintenance means observation.

From Nov. 2007 to Jan. 2008, we invited 4 people from Mongolian side here for technology exchange and training. During training process, they had systematically learned courses of elementary knowledge about historical building protection, maintenance specifications, protection of mural and color decoration, etc. Finally, they were awarded the Training Certificates. Through training and practice work of cultural heritage conservation and restoration, Mongolian technicians grasped the ideas, means and technology of historical building maintenance by the large, which played an important role in follow-up protection of the front square maintenance project of Mongolia Bogd Khaan Palace Museum.

G.Enkhbat, the Director of Mongolian Protection Center of Cultural Heritage, with four others, came to Shaanxi Institute for the Preservation of Cultural Heritage for study visit from May 22nd to May 24th in 2011. During the visit, the two sides held discussions, Director G.Enkhbat expressed the purpose of this visit: Firstly, to study the organs, staff and equipments of related cultural heritage protection organizations in China; Secondly, to seek cooperation, with the hope of establishing teamwork with Shaanxi Institute for the Preservation of Cultural Heritage on temple colored painting, mural restoration, etc. Later, Director G.Enkhbat made a field trip not only to the laboratory and restoration room of Shaanxi Institute for the Preservation of Cultural Heritage of Cultural Property but also to the mural protection and restoration site.

蒙古国专业人员现场交流
Mongolia Professionals making Field Exchange

蒙古国专业人员现场实践
Mongolia Professionals making Field Practice

德国专家现场交流
German Experts making Field Exchange

美国专家现场交流
U.S. Experts making Field Exchange

蒙古国宗教人士现场交流
Mongolia Religionists making Field Exchange

国外专业人员交流
Foreign Professionals Exchanging Activity

Дугаар 8020

Yang Bo
Хүндэт .... Таны

Соёл, урлагийг хөгжүүлэх үйл хэрэгт оруулсан хувь нэмрийг тань үнэлж Монгол Улсын Боловсрол, соёл, шинжлэх ухааны яамны "ЖУУХ БИЧИГ"-ээр шагнаж үнэмлэх олгов.

Монгол Улсын Боловсрол, соёл, шинжлэх ухааны сайд /О.Энхтүвшин/

2007 оны 10 сарын 05-ны өдөр
Улаанбаатар хот

荣誉证书　编号：8020
尊敬的 杨博先生
对于你为保护蒙古国文化遗产所做出的工作，特颁发此证书。
蒙古国文化科技教育部 部长 ： O. Enkhtuvshin
2007年10月5日

Certificate of Honor　No.:8020
Respected Mr. Yang Bo
For the contribution you have done to protect the cultural heritage in Mongolia, we hereby deliver this Certificate to you.
Director of Mongolia Ministry of Education and Culture:
O. Enkhtuvshin
Oct. 5th 2007

Дугаар 8021

Ma Tu
Хүндэт .... Таны

Соёл, урлагийг хөгжүүлэх үйл хэрэгт оруулсан хувь нэмрийг тань үнэлж Монгол Улсын Боловсрол, соёл, шинжлэх ухааны яамны "ЖУУХ БИЧИГ"-ээр шагнаж үнэмлэх олгов.

Монгол Улсын Боловсрол, соёл, шинжлэх ухааны сайд /О.Энхтүвшин/

2007 оны 10 сарын 05-ны өдөр
Улаанбаатар хот

荣誉证书　编号：8021
尊敬的 马途先生
对于你为保护蒙古国文化遗产所做出的工作，特颁发此证书。
蒙古国文化科技教育部 部长 ： O. Enkhtuvshin
2007年10月5日

Certificate of Honor　No.:8021
Respected Mr. Ma Tu
For the contribution you have done to protect the cultural heritage in Mongolia, we hereby deliver this Certificate to you.
Director of Mongolia Ministry of Education and Culture:
O. Enkhtuvshin
Oct. 5th 2007

蒙古国业务人员在西安参观学习
Mongolia Personnel making study visit in Xi' an

蒙古国官员在西安开展学习交流活动
Mongolia officials making Learning and Exchanging Activities in Xi' an

培训合影
Photo of Training Group

颁发证书
Photo of Awarding Certificates

2.回访工作

2.1 回访基本情况

2010年12月6日至12月11日回访期间，按照工作计划，援蒙项目回访工作组首先对蒙古国博格达汗宫博物馆门前区保护工程进行回访，与蒙古国教育文化科技部官员、博格达汗宫博物馆馆长、蒙古国文化遗产保护中心主任进行了工作会谈和现场勘察；其次与蒙古国教育文化科技部、蒙古国文化遗产保护中心、蒙古国国立历史博物馆、艺术博物馆、自然博物馆相关人员进行了文化交流活动。双方希望继续在技术交流、人员培训、保护项目等方面开展进一步的合作。

2.2 现场核查情况

陕西省文物保护研究院援蒙项目回访工作组对博格达汗宫博物馆门前区大门、牌楼、照壁、旗杆等十个单体分别从基础、结构、木构件表面的地仗、油饰及彩画等方面进行了认真、仔细的现场保存现状核查。通过核查表明，在历经蒙古国三年多极端气候的考验之后，蒙古国博格达汗宫博物馆门前区古建筑保存现状总体良好。古建筑基础和结构方面的维修效果良好，没有出现任何问题。古建油饰彩画的保存状况理想，完全符合设计要求。木构件表面的地仗、油饰保存情况差别较大，暴露在阳光直射的区域出现不同程度的褪色、地仗龟裂、起甲及剥离，其他区域保存状况良好。

现场勘察
Field Inspection

2. Revisit and Review work

2.1 Revisit overview

During revisit time from Dec. 6th to Dec. 11th in 2010, revisit assistance group to Mongolia first revisited the front square maintenance project of Mongolia Bogd Khaan Palace Museum, holding talks and field investigation with officials from Mongolia Ministry of Education and Culture, curator of Mongolia Bogd Khaan Palace Museum and director of Mongolian Protection Center of Cultural

Heritage, then conducted cultural exchange activities with related person from Mongolian Ministry of Education and Culture, Mongolian Protection Center of Cultural Heritage, National Museum of History, Museum of Art, Museum of Natural History. Further cooperation in technology exchange, staff training and protection project would also be welcomed by the two nations.

2.2 Field inspection on assistance project to Mongolia

Assistance group from Shaanxi Institute for the Preservation of Cultural Heritage carefully inspected ten sections including the front gate, decorated archways, entrance screen wall, flagpole, etc. of the front square of Bogd Khaan Palace Museum from several aspects such as foundation, structure, base layer, paint coat decoration and colored painting on the surface of wooden components. The inspection result indicated that the conservation state of ancient architectures at the front square of Bogd Khaan Palace Museum was basically fine after three years' baptism of extreme climate in Mongolia. The maintenance effect of foundation and structure of ancient architectures looked good without any problem, so did paint coat decorations and colored paintings on the surface of ancient architecture, which were completely compliance with design requirements. The conservation state of base, paint coat decorations on the surface of wooden components varied considerably, the area exposed to sunlight was found to be color faded, base cracked, scabbed and peeled while other area looked good.

回访现状照片

Photos of ancient architecture at the front square of Bogd Khaan Palace Museum

大门现状
Condition of the Front Gate

中牌楼现状
Condition of middle Archway

东牌楼现状
Condition of east Archway

西牌楼现状
Condition of west Archway

旗杆现状
Condition of Flagpole

西便门现状
Condition of west Side Door

东便门现状

Condition of east Side Door

南宫墙现状

Condition of south Palace Wall

栅栏墙现状
Condition of Stockade Wall

照壁现状
Condition of Entrance Screen wall

博格达汗宫博物馆门前区现状

The front square of Bogd Khaan Palace Museum

2.3 当地气候环境的影响及维护保养建议

气候环境的影响

蒙古国地处内陆，大部分地区为山地或高原，平均海拔1600米，自然地理状况对其气候有非常大的影响，其气候具有强烈的大陆性特征，季温差和日温差均很大。乌兰巴托位于蒙古高原中部，海拔1351米，属典型的大陆性气候，冬季最低气温达－40℃，夏季最高气温达35℃，年平均气温－2.9℃。季节变化明显，冬季长（十一月至四月），天寒地冻，降雪时间长。夏季短（七、八月），昼夜温差大，日平均最高温度与最低温度的温差达到25℃以上。季节温差大，冬夏之间最大温差达到75℃。春秋季节天气变化剧烈，九、十月份一天中天气剧变时甚至降雪。全年多风沙，降水很少，年平均降水量约120－250毫米，70％集中在七、八月。一年四季平均日照时间长，紫外线强。这些外界因素对木结构的古建筑长久保存造成了极大的影响。温度的影响主要表现在高温和温度的急遽变化两个方面。温度过高会使氧化分解等化学反应加速进行，因此温度越高，木材等有机材料变质老化的速度越快；而温度的急遽变化产生的反复热胀冷缩过程极易使文物材质弹性疲乏，镶嵌装饰或接合部位容易松脱。阳光直接照射，紫外线强，因紫外线所具有的能量足以破坏聚合物的化学键，通过光化学氧化反应造成老化降解破坏古建筑表面油饰及彩画。全年多风沙也对古建筑油饰及彩画表面造成磨蚀的损毁作用。

维护保养建议

蒙古国博格达汗宫博物馆自从援蒙项目结束后，没有采取有效的方法对古建筑进行日常的维护、保养，如对残损的表面油饰也没有及时进行修补等日常维护工作。同时，博物馆门前区积雪堆积、杂草丛生，也不利于古建筑的长久保存。

鉴于此，西安文物保护修复中心援蒙项目回访工作组对蒙方做好古建筑的日常维护和保养提出了以下建议：

建议蒙方在博格达汗宫博物馆古建筑群内建立文物保护环境监测系统，做好日常数据的记录，通过数据积累及时总结和发现对古建筑保存影响的因素，周期为1－2年。同时建议，聘请专业勘察机构定期对古建筑大木构件进行结构稳定性检测，检测周期以7－10年为宜。

建议做好古建筑的日常保养，及时清理地面及台明的积雪、杂草，对于大门滴水处进行苫护，防止积雪堆积、杂草丛生、反复冻融等现象，以避免砖体出现冻融开裂等病害。清理卫生时注意不要用水冲洗古建筑的构件、窗花、装饰物。

鉴于蒙古国极端的自然气候环境，建议将博格达汗宫古建群的油饰彩画维护保养周期调整为3年（中国国内古建筑油饰彩画的正常维护保养周期为5—10年）。在维护保养中，应当对大木构件地仗的缺损部位进行修补，并对古建筑表面进行重新油饰。

建议博格达汗宫古建筑的铁皮瓦屋面维护保养周期为5年，维护保养中主要做好铁皮瓦屋面的物理除锈、防锈及表面重新油漆工作。

建议做好彩画部分的表面清洁，特别是防止鸟粪、灰尘等的污染。对于鸟雀能容身的部位建议进行拉网防护。

2.3 Impact of local climate and advices for maintenance

Impact of local climate

Mongolia is located inland, most regions are mountains and plateaus with average altitude of 1600 meters, natural geographical condition has great effect on local climate which possesses continental characteristic, the season and daily temperature difference is huge. Ulaanbaatar sits in the middle of Mongolia with altitude of 1351 meters, the climate there carries continental characteristic, the minimum temperature in winter is minus 40 degrees while the maximum point in summer hits 35 degrees, annual average temperature is minus 2.9 degrees. Seasonal variation there is obvious, winter time is long (from Nov. to Apr.), so is the time of snowfall. Summer time is short (Jul. and Aug.) with big temperature difference between day and night which reaches over 25 degrees between maximun and minimun in a day. There is big temperature difference between seasons which reaches 75 degrees at most. The weather changes severely in spring and autumn, snowfalls even can be seen in September and October. Sand wind often blows all the year round with less annual rainfall of average 120-250 mm, 70 percent amount of rainfall intensively happens in July and August. The average sunshine time is long throughout the year with strong ultraviolet rays. These climatic factors above have a strong impact on the conservation of ancient architectures. The impact of temperature mainly includes two aspects of high temperature and sudden change of temperature. High temperature would accelerate oxygenolysis which speeds up the metamorphism and aging pace of organic materials such as timber, while repeating expansion and contraction due to the sudden change of temperature would easily deteriorate the material quality of cultural heritage and loosen the inlay decoration and junction parts. The power of strong ultraviolet rays is strong enough to destroy the chemical bond of polymer, paint coat decorations and colored drawings on the surface of ancient architectures can be not only damaged by aging and degradation resulted in actinology oxidation, but also be destroyed by sand wind.

Advices for maintenance

After the assistance project for Mongolia Bogd Khaan Palace Museum, there are no effective measures being taken to maintain and preserve the ancient architectures, for instance, damaged surface painting decorations doesn’ t get timely repair. Meanwhile, snow cover and weeds at the front square of Museum can not serve the purpose of long-term conservation of ancient architectures. For that reason, revisit assistance group from Xi'an Centre for the Conservation and Restoration of Cultural Heritage to Mongolia would like to give several advices as below to Mongolia side for daily maintenance of ancient architectures:

Environmental monitoring system for cultural heritage protection should be set up inside the ancient architectural complex at Bogd Khaan Palace Museum, daily data must be recorded carefully with the cycle period of 1-2 years so as to timely summarize and discover the factors affecting the conservation of ancient architectures. Besides, exploration agency should be employed to periodically detect structural stability of wooden components of ancient architectures with the cycle of 7-10 years.

Daily maintenance for ancient architectures should be done attentively, snow cover and weeds on the ground and platform should be cleaned promptly. For the sake of keeping off snow cover, weeds, repeated freeze and melt on the ground where dripping, cogongrass covering must be capped on it to avoid brick cracking. Do not clean the components, window grille and ornaments of ancient architectures with water when scavenging.

Considering the extreme climate condition in Mongolia, the maintenance period for colored painting decorations on ancient architectures should be switched to 3 years (this periodic time in China is 5-10 years normally). The absence and damaged section of the base layer on wooden components must be repaired when maintaining, the surface of ancient architectures should be also repainted.
The periodic time of maintenance for iron sheet and tile roof on ancient architectures should be 5 years, physical derusting, rust protection and repainting job must be stressed when maintaining.

Surface cleaning of colored painting should come into notice, especially to avoid the contamination of birds droppings, dust, etc. For the place where birds may inhabit, protective netting must be introduced.

大事记

2004年

4月：文化部孙家正部长、国家文物局单霁翔局长率中国政府文化代表团访问蒙古国。蒙古国总理在会见代表团时表示希望中蒙两国在文物保护考古等方面进行合作。

7月：国家文物局童明康副局长率文物代表团访问蒙古国，双方就维修博格达汗宫博物馆大门、合作考古及举办文物展等事宜达成共识。

2005年

5月27日至6月7日：国家文物局组织专家赴蒙古国对博格达汗宫博物馆门前区进行了实地勘察。

7月7日：国家文物局发出《关于蒙古国博格达汗宫博物馆门前区维修保护初步设计方案的批复》，批复同意了陕西省文物保护研究院制订的《蒙古国博格达汗宫博物馆古建筑维修设计方案》。

2006年

2月27日：国家文物局致函陕西省文物局，委托陕西省文物保护研究院承担蒙古博格达汗宫门前区修复工程，并请省局协调做好准备工作。

4月6日：国家文物局致函陕西省文物保护研究院，委托其实施蒙古国博格达汗宫门前区维修保护工程项目。

5月27日：举行工程开工典礼，蒙古国总理米耶贡布・恩和包勒德、教育文化科技部部长恩赫图布欣和中国驻蒙大使高树贸、国家文物局副局长董保华等出席了开工典礼。中央电视台和蒙方媒体进行了报道。

5月27日至9月20日：陕西省文物保护研究院开展第一期工程。

9月13日至18日：国家文物局组织国内相关专家检查维修工地,通过了第一期工程的阶段性验收。

9月20日：中国驻蒙使馆举行记者招待会，蒙古国多家电视广播媒体和平面媒体记者参加招待会，并进行正面报道，给予积极评价。乌兰巴托电视台还做了专访节目。

2007年

4月25日：工程二期开工。

5月14日至18日：国家文物局派出王军、许言、凌明、饶权、李永革、王效清6人，对维修工程进行质量检查。

9月27日至30日：国家文物局专家与蒙方专家组成专家组，对工程验收。经过联合验收，专家组认为该工程工程质量和效果良好，一致通过验收。

10月8日：举行工程竣工典礼，蒙古国教育文化科技部部长恩赫图布欣、中国驻蒙大使余洪耀、中国国家文物局副局长张柏等出席了竣工典礼。

2010年

12月6日至11日，根据国家文物局的指示，应蒙古国教育文化科技部的邀请，陕西省文物保护研究院张颖岚主任、马琳燕副研究员、杨博工程师等一行3人赴蒙古国乌兰巴托，对援蒙项目进行了回访。

Chronicle of Events

The year 2004

April: Sun Jiazheng, minister of Chinese Ministry of Culture, Shan Jixiang, director of State Administration of Cultural Heritage, with cultural delegation from Chinese government, visited Mongolia Republic. Mongolia Prime Minister expressed his hope of Sino-Mongolia cooperation on cultural heritage protection and archaeology when meeting with Chinese delegation.

July: Tong Mingkang, deputy director of State Administration of Cultural Heritage, with cultural heritage delegation, visited Mongolia and reach a consensus with Mongolia side on the front square maintenance project of Bogd Khaan Palace Museum, teamwork archaeology and cultural heritage exhibition, etc.

The year 2005

May 27th to June 7th : State Administration of Cultural Heritage specially commissioned Shaanxi Institute for the Preservation of Cultural Heritage to form a experts group for field investigation and mapping at the front square of Bogd Khaan Palace Museum.

July 7th : Chinese State Administration of Cultural Heritage issued "Reply on preliminary design scheme for the front square maintenance project of Bogd Khaan Palace Museum" which approved "Design scheme for the front square maintenance project of Bogd Khaan Palace Museum" made by Shaanxi Institute for the Preservation of Cultural Heritage.

The year 2006

February 27th : Chinese State Administration of Cultural Heritage wrote to Shaanxi Provincial Bureau of Cultural Heritage and commissioned Xi'an Centre for the Conservation and Restoration of Cultural Heritage to undertake the front square maintenance project of Bogd Khaan Palace Museum, Shaanxi Provincial Bureau of Cultural Heritage was required to make preparation work.

April 6th : Shaanxi Institute for the Preservation of Cultural Heritage was specially commissioned by State Administration of Cultural Heritage to implement the front square maintenance project of Bogd Khaan Palace Museum.

May 27th : Project commencement ceremony was held, Miyeegombo Enkhbold, prime minister of Mongolia, Enhertubexin, minister of Mongolia Ministry of Education and Culture, Gao Shumao, Chinese ambassador in Mongolia, Dong Baohua, deputy director of Chinese State Administration of Cultural Heritage, etc. attended the commencement ceremony. CCTV and Mongolia media reported this event.

May 27th to September 20th : Shaanxi Institute for the Preservation of Cultural Heritage carried out the first-stage project.

September 13th to 18th : Chinese State Administration of Cultural Heritage organized domestic experts to inspect the project site, the first-stage project passed the acceptance check.

September 20th : Chinese embassy in Mongolia held press conference, Mongolian TV media, broadcast media and print media attended this conference and reported positively with highly appraisal. Ulaanbaatar TV conducted exclusive interview.

The year 2007

April 25th : The second-stage project came into operation.

May 14th to 18th : Chinese State Administration of Cultural Heritage sent out 6 people named Wang Jun, Xu Yan, Rao Quan, Li Yongge and Wang Xiaoqing to inspect the project quality.

September 27th to 30th : The experts, sent out by Chinese State Administration of Cultural Heritage, along with experts from Mongolia conducted project acceptance check . All experts considered this project to be qualified in quality and fine in effect, the project got unanimous approval.

October 8th : Construction completion ceremony was held, Enhertubexin, minister of Mongolia Ministry of Education and Culture, Yu Hongyao, Chinese ambassador in Mongolia, Zhang Bo, deputy director of Chinese State Administration of Cultural Heritage, etc. attended the completion ceremony.

The year 2010

December 6th to 11th : Zhang Yinglan, director of Shaanxi Institute for the Preservation of Cultural Heritage， along with Ma Linyan, associate researcher, and Yang Bo, engineer, revisited Ulaanbaatar Mongolia for this assistance project.

后 记

陕西省文物工作者承担的无偿援助蒙古国文化遗产保护项目完成后，在上级主管部门的支持下，在单位各部门同仁的努力下，《博格达汗宫博物馆维修工程》一书得以出版。

在项目施工和报告出版过程中得到中国驻蒙大使馆、国家文物局、陕西省文物局、蒙古国文化教育科技部、内蒙古二连浩特市恐龙博物馆、博格达汗宫博物馆、文物出版社等单位的大力支持与帮助，使本书能够得以顺利出版，使无偿援助项目过程中付出的汗水以及文物保护中的先进技术和科研成果得以展示，表示感谢！

同时对于陕西省文物保护研究院所有参与项目与编辑本书的同志，表示诚挚的感谢！

编　者

2013年6月

Afterword

Following the completion on the gratis assistance over Mongolia' s cultural heritage protection undertook by staffs from Shaanxi Province, the book entitled 《The Maintenance Project of Bogd Khaan Palace Museum》 finally and officially comes to the public under the joint efforts from superior authorities and relevant departments.

During the course of this project, the Chinese Embassy in Mongolia, State Administration of Cultural Heritage, Shaanxi Provincial Bureau of Cultural Heritage, Mongolian Ministry of Education and Culture, Inner Mongolia Erenhot City Dinosaur Museum, Bogd Khaan Palace Museum, Relic Press, etc. generously provided their abundant supports and helps, which also contributed to the successful publication of this book, this is a book which records every perspiration during implementing this project, and displays the advanced technics in protecting the cultural relics.

Meanwhile, sincere gratitude is also extended hereby to those from Shaanxi Institute for the Preservation of Cultural Heritage who participated in this project and book-publishing process.

June 2013

策 划
张颖岚
撰 稿
杨 博 马琳燕 齐 扬 张 炜 杨秋颖 马 途
摄 影
杨 博 马 途 何永斌
版式设计
李 清
翻 译
杜 娟

图书在版编目（CIP）数据

博格达汗宫博物馆维修工程 / 陕西省文物保护研究院著. -- 北京 : 文物出版社, 2014.6
ISBN 978-7-5010-4011-7
Ⅰ. ①博… Ⅱ. ①陕… Ⅲ. ①博物馆－维修－乌兰巴托 Ⅳ. ①TU746.3
中国版本图书馆CIP数据核字(2014)第101097号

博格达汗宫博物馆维修工程

著　　者 陕西省文物保护研究院
责任编辑 李 睿
责任印制 陆 联
责任校对 李 薇 陈 婧
出版发行 文物出版社
地　　址 北京市东直门内北小街2号楼
邮政编码 100007
网　　址 http://www.wenwu.com
邮　　箱 web@wenwu.com

制版印刷 北京雅昌艺术印刷有限公司
经　　销 新华书店
版　　次 2014年6月第1版第1次印刷
开　　本 787×1092　1/8
印　　张 26
书　　号 ISBN 978-7-5010-4011-7
定　　价 500.00元